The Art of High Performance Computing
for Computational Science, Vol. 1

Masaaki Geshi
Editor

The Art of High Performance Computing for Computational Science, Vol. 1

Techniques of Speedup and Parallelization for General Purposes

 Springer

Editor
Masaaki Geshi
Osaka University
Toyonaka, Osaka, Japan

ISBN 978-981-13-6196-8   ISBN 978-981-13-6194-4   (eBook)
https://doi.org/10.1007/978-981-13-6194-4

Library of Congress Control Number: 2019934788

This Springer imprint is published by the registered company Springer Nature Singapore Pte Ltd.
The registered company address is: 152 Beach Road, #21-01/04 Gateway East, Singapore 189721, Singapore

Preface

Supercomputers have been used in various fields since around the 1980s. At that time, a vector architecture was the mainstream of the supercomputer, and program developers worked on vector tuning. However, in the 1990s, a parallel architecture emerged and the mainstream of the supercomputer changed. The development of supercomputers is ongoing under this trend. Currently, in the era of massively parallel or manycore systems, in the top 500 ranking in the competition of supercomputers for LINPACK performance, some of the top ten computers of the November 2018 ranking have over 1 million cores, and one even has over 10 million cores.

This is not just a pleasure for users of such supercomputers. Different from the vector tuning in the vector computer, the performance improvement cannot be expected simply for taking the vector length as long as possible. The performance will not improve unless you are familiar with the characteristics of the hardware such as knowing the size of a cache and tuning programs so that the proper amount of data is placed in the cache, concealing communication latency, and reducing the number of data communications. This is the current supercomputer. A hotspot depends mainly on algorithms of calculation methods that are distinguished in the field, and there is no common development policy. The supercomputer with tens of millions of cores has already appeared and more will be created in the future. There is also a problem such as whether a parallelization axis large enough to exhaust all the cores is in the algorithm. If not, we have to create it or consider an alternative.

In addition, although the correspondence to machines equipped with accelerators has advanced in recent years, there are some fields where performance improvement has not yet sufficiently advanced. In some cases, it may be necessary to tackle improvements in fundamental algorithms.

Only computer scientists and some computational scientists took the above-mentioned steps. However, if you would like to use the current supercomputers effectively, this knowledge and these techniques are necessary for more users. Such knowledge is indispensable for researchers who are developing software, and there is an increasing number of things to know from the user's point of view. Even if we compile and execute a software program on a supercomputer without thinking about hardware, we will not be able to obtain better performance than PC clusters widely used at the laboratory level.

Although the tuning techniques become more difficult year by year, the fact is that we can obtain excellent performance if we simply tune the software program properly. If we make a supercomputer, we must have human resources with the techniques to master it.

In some fields, it is difficult to recognize the speedup of a program as a significant result. In some cases, software development itself is not recognized as an important achievement. For researchers who are promoting software development, therefore, the speedup of the program and/or the software development includes a great risk of being unable to produce research results during the given term. However, it is the essence of research if results can be obtained when otherwise they could not be achieved without a highly tuned program in realistic computing time—unless a huge amount of computational time were spent. Therefore, it is extremely important to decrease the computational time as much as possible.

For the general public, there are only reports of alternate hopes and despair in the top 500 rankings, but for those in computational science, the peak performance of such machines and the benchmark performance of LINPACK are merely for reference. The important point is whether we can achieve the best performance by tuning the software program we create or use to obtain scientific results. Only such software programs can demonstrate the real value of supercomputers.

This series of books is written about the basics of parallelization, the foundation of numerical analysis, and related techniques. Even if it is mentioned as a foundation, we assume the reader is not a complete novice in this field; so if you would like to understand programming from the beginning, you can learn the basics from another book, more suitable for that purpose. Our readers are assumed to be those who have an understanding of physics, chemistry, and biology, as well as those in the fields of earth sciences, space science, meteorology, disaster prevention, and manufacturing, among others. Furthermore, we assume readers to be those who use numerical calculation and simulation as research methods. In particular, we assume them to be those who develop software code.

Volume 1 includes field-independent general numerical analysis and parallelization techniques. From Chaps. 1 to 5, the basic techniques of speedup and parallelization are provided, including an explanation of recent hardware. In Chap. 6, the basics and practice of the linear algebra calculation libraries BLAS and LAPACK are introduced. In Chap. 7, high-performance algorithms for numerical linear algebra are provided. Chapter 8 presents fast Fourier transform (FFT) in large-scale systems. Chapter 9 discusses optimization and related topics such as debug methods and version control systems, and in Chap. 10, techniques concerning computation accuracy are introduced. Although several examples of methods of materials science are used, most techniques can be applied to other fields.

Chapters 1–5 are useful for readers who would like to have an overview of high-performance computing (HPC) from the point of view of programming with message passing interface (MPI) and OpenMP. Chapter 6 is useful for readers who would like to learn the details of BLAS and LAPACK. This chapter introduces some examples that can be easily reproduced using free software. Chapter 7 focuses on sparse matrix calculations that are different from those in Chap. 6, which deals with dense matrix calculations. Chapter 8 focuses on FFT calculations from the basics. Chapter 9 introduces the basics of programing such as creating a program without bugs, how to debug, and how to use useful tools. Chapter 10 is different from the preceding chapters and includes content found in few other books. Almost all HPC-related books focus on speedup of programs; however, it is essential that the calculation results are numerically correct. There are a number of things to keep in mind in order to guarantee calculation accuracy. They are introduced in this chapter, thus the content of Chap. 10 is recommended to researchers for all fields.

Volume 2 includes advanced techniques based on concrete applications of software for several fields, in particular, the field of materials science. From Chaps. 1 to 3, advanced techniques are introduced by using the tuning results executed on the K computer. The authors provide several examples from various fields. In Chap. 4, the order-N method based on density functional theory (DFT) calculation is presented. Chapter 5 introduces acceleration techniques of classical molecular dynamics (MD) simulations. In Chap. 6, techniques for large-scale quantum chemical calculation are given.

This book is revised and updated from the Japanese volume published by Osaka University Press in 2017, *The Art of High Performance Computing for Computational Science, Vol. 1* (Masaaki Geshi, ed., 2017). That book was based on the lectures "Advanced Computational Science A" and "Advanced Computational Science B", broadcast to a maximum of 17 campuses through videoconference systems from 2013. All the texts and videos were published on websites (only in Japanese). These were parts of the human resource development programs that we tackled as part of the Computational Materials Science Initiative (CMSI) project organized by the Ministry of Education, Culture, Sports, Science and Technology (SPIRE) field 2 (New materials/energy creation), the so-called K computer project. These lectures aimed to contribute to developing young human resources, centering on basic techniques that will not change for a long time even though it is a computer that progresses day by day. These lectures were offered from the Institute for Nano Science Design, Osaka University. We gathered up to about 150 participants per lecture, and participants exceeded a total of 6500 people in the past 6 years. The videos of the lectures and lecture materials are open to the public on the Internet, and anyone can learn from their content at any time in Japanese. These lectures continue to be distributed with slight changes in the organizational structure even after the project has ended.

I would like to express my deep appreciation as editor to the authors, who cooperated in the writing of the series of books. I would also like to thank the staff of Springer for publishing the English version. I believe that the techniques cultivated in the project of the K computer in Japan contain many useful contents for future HPC. I hope this knowledge will be shared throughout the world and will contribute to the development of HPC and the development of science in general.

Osaka, Japan   Masaaki Geshi
November 2018

Contents

Chapter 1
High-Performance Computing Basics

Takahiro Katagiri

Abstract This chapter explains the basics of speedup programs with simple examples for numerical computing. Parallel processing as well as the trends in computer hardware are explained in detail to understand how high-performance computing works. Several key technologies for code tuning, such as pipelining, cache optimizations, and numerical libraries, are also explained using sample programs from basic numerical computations.

1.1 Trends in Recent Computer Architectures

It is important to know the trends in recent computer architectures to establish speedup programs. Conventional technologies of speedup may be effective because of the progress in computer architectures. On the other hand, it is also possible for ineffective technologies for past computer architectures to become effective in recent computer architectures. For this reason, we summarize the nature of current computer architectures in the first part of this chapter.

The classifications of current computer architectures are summarized as follows:

- **Multicore Type**: In this type of computer architecture, the frequency of logic gates inside the computer is lowered to reduce electrical power while the number of computational units, called "cores", is increased to obtain high-performance computations. The number is around 10 cores in commodity types of CPUs, and it establishes around 30 cores in high-end type of CPUs, as of 2017.
- **Many-Core Type**: This type of computer architecture is similar to a multicore where the frequency is designed to be lower, but with a greater number of cores. The number of cores is around 60 physical cores, and the actual parallelism is more than 200, as of 2017.
- **Accelerator Type**: In addition to CPU, accelerators are used in this type of architecture. A Graphics Processing Unit (GPU) is commonly used as an accelerator.

T. Katagiri (✉)
Information Technology Center, Nagoya University, Nagoya, Japan
e-mail: katagiri@cc.nagoya-u.ac.jp

© Springer Nature Singapore Pte Ltd. 2019
M. Geshi (ed.), *The Art of High Performance Computing for Computational Science, Vol. 1*, https://doi.org/10.1007/978-981-13-6194-4_1

In this chapter, we will explain technologies of speedup by CPUs for multicore and many-core architectures. The effective techniques for both types are different in some aspects, but same techniques can be shared in many cases. Execution condition affects the performance of a code; thus, in applying the techniques shown in this book, execution conditions, such as the number of cores used and/or the size of problem, should be taken into consideration to determine the best technique for speedups.

1.2 Computer Architecture

In this section, we will explain the general aspects of speedup programs on recent computer architectures.

1.2.1 Pipelining

One of the important characteristics of current computer architectures is the implementation of pipelining. Pipelining is a method to obtain high efficiency for processes inside computers. It divides continuous processes into several stages and performs parallel computations to obtain high efficiency.

Now, we consider applying the pipelining to process a numerical computation. Let a dense matrix be $A \in \mathfrak{R}^{n \times n}$, a dense vector be $x \in \mathfrak{R}^n$, and a result of dense vector be $y \in \mathfrak{R}^n$. With these notations, we compute the following:

$$y = Ax. \tag{1.1}$$

To perform the first computation in Eq. (1.1), we need to obtain the elements of $A_{1,1}$ for the matrix A and the elements of x_1 for the vector x. Let the computer language be Fortran to describe the computation in Eq. (1.1). We use a two-dimensional array of **A(n, n)** for the matrix A and one-dimensional arrays of **x(n)** and **y(n)** for the vectors x and y. With these notations, we show an overview of the pipelining in Fig. 1.1.

Figure 1.1a processes are divided into four stages: (1) Fetch **A(i, j)** from memory, (2) Fetch **x(i)** from memory, (3) Perform the computation **y(j) + A(i, j) * x(i)**, and (4) Store **y(j)** to memory. In the process shown in Fig. 1.1a, the next computation cannot execute until finishing the fourth stage. The only time a computational unit is used is in stage (3). Let the time of each stage be 1 time unit. In this case, the computational unit is not working in three of four stages. With this, the efficiency of the computational unit is poor with only 25% efficiency.

Pipelinization is applied to improve the computational efficiency as compared to sequential computing shown by Fig. 1.1a, b. To perform pipelinization, the hardware is modified to do parallel processing of each stage. For example, after finishing stage

Example of Computation Unit

▸ **Ex.: Process of Computation Unit**

Note: The process is different from actual computation units in CPU.

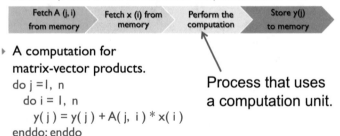

▸ **A computation for matrix-vector products.**

do j = 1, n
  do i = 1, n
    y(j) = y(j) + A(j, i) * x(i)
enddo; enddo

Process that uses a computation unit.

▸ **If pipelinization is not adapted, the following inefficient process are performed.**

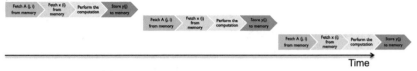

Time

(a) A Case of Sequential Computing.

Example of Computation Unit

▸ It is inefficient because only a time unit is used during four time units. Efficiency of the computations is 1/4 = 25%.

▸ The following pipelinization shows every computations per time unit when enough time is spent. Efficiency of the computation is 100%.

> Enough time" in this example is the optimum number of loop iterations. Bigger size of N gives us high efficiency of pipelinization.
> The efficiency of computation goes up.
> > When N is small, the efficiency of computations is basically low.

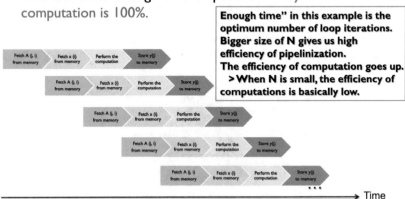

Time

(b) A Case of Pipelinization.

Fig. 1.1 A pipelinization in matrix–matrix multiplications

(1) Fetch **A(i, j)** from the memory, the circuit lines of fetching data from the memory are not being used, and hence the next stage can start by using the circuit lines to process (2) Fetch **x(i)** from memory. By using this framework of computations, all four stages are processing when enough time is spent. "Enough time" here is the optimum time to do the pipelining. Therefore, the efficiency of using a computational unit reaches to near 100%. This is the nature of pipelinization.

Take note that the number of matrix **n**, which is for matrix–vector multiplication, needs to be large if you want to obtain high efficiency.

Generally speaking, it requires having a large length of loops to obtain high efficiency for pipelining. The time to obtain enough high efficiency for the pipelining is called **pipeline latency**. It is better for the pipeline latency to be small. To obtain high efficiency in a machine with a large pipeline latency, larger length of loop is desirable.

1.2.2   Hierarchical Memory

Another characteristic for recent computer architectures is the memory which consists of hierarchical organizations. We call this memory organization "Hierarchical Memory". There are memories with small capacities but with high access speed as well as memories with large capacities but with low access speed. It is impossible to make a memory with a high capacity and high access speed in from the point of view of engineering. Hence, to reduce memory cost and to establish high access speed, memories should be organized in the form of hierarchical memory. Figure 1.2 shows the hierarchical memory.

Figure 1.2 shows the organization of memories where registers, caches, main memory, and hard disk are arranged in the order of decreasing memory speed. Recent computer architectures are organized with caches consisting of Level 1 (L1) cache, which means independent cache per core, and with a shared cache between several cores. For the shared caches, it takes two or more hierarchies in current computer

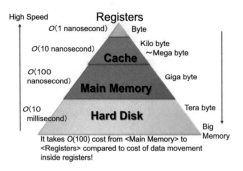

Fig. 1.2 Organization of hierarchical memory

architectures. With respect to the trends in computer architecture, it goes up to four or more levels for the shared caches. In particular, we call the nearest cache from the main memory or the farthest cache from cores as **Last Level Cache (LLC)**.

In Fig. 1.2, we can imagine that 10 times speedups can be established to store data in cache as compared to storing data in the main memory, where it is normally stored, to reduce time for data movement. In addition, if we can store data in registers in the program, it can reach 100 times speedups as compared to the data that is usually stored in the main memory.

As we have discussed above, hierarchical memory is an important factor to be considered in obtaining high performance in current computer architectures.

1.2.3 Heterogeneousness

To improve the computational performance per watt, different kinds of computational units are implemented in the current computer architectures. For example, sequential part should be executed on CPUs with high frequency and parallel parts should be executed on CPUs with low frequency in parallel. This organization of architectures is called **heterogeneous organization**.

To make computers with heterogeneous organization, one of the effective approaches is using accelerators as we explained in the beginning of this chapter, since computer with accelerators needs CPU to execute programs. Usually, CPU has a different architecture as accelerators. Hence, this approach provides us **heterogeneous computations**. Reducing communications (data movement) between the main memory and the device memory is a crucial factor to speed up in the accelerators type. This will be further explained in Chap. 3.

On the other hand, several CPUs are implemented in a node. In this environment, access speed in memory is different among CPUs to reduce difficulty of design and power budget. This access is called "**heterogeneous memory access.**" This organization is called **Nonuniform Memory Access (NUMA)**. It is a crucial factor to allocate arrays near the memory to speed up for the NUMA organization. This will also be further explained in Chap. 3.

There is another organization where single computations are faster than double computations. This organization provides "**heterogeneous computation**" in terms of computational speed. Utilizing **mixed precision computations** [1, 2] is a considerable approach for heterogeneous computation.

As the author mentioned above, current computer architectures have several kinds of heterogeneity. To establish speedup, we should know the nature of the heterogeneity on target computer architectures.

1.2.4 Parallelism

At the last part of the section, the author explains one of most important features for recent computer architectures called **parallelism**. As the author mentioned, 200 parallelisms or more can be utilized for current computer architectures as of 2017. With respect to the trends in computer architectures, it is expected that several thousands of parallelisms will be available in the near future.

To utilize such enormous parallelisms, several technologies are being studied. Definitely, parallelism in programs is important and required, but it is not enough. Increasing parallelism for pipelining such as implementation to increase the loop length needs to be considered. In addition, implementations with respect to machine codes which use **instruction-level parallelism** are also required, such as **Single Instruction Multiple Data stream (SIMD)** operations.

1.3 Techniques to Speedup

In this section, general techniques of code optimization to speedup programs are explained.

1.3.1 Continuous Loop Access

First of all, we treat many arrays in the program. Hence, we need to consider access patterns for the arrays to perform "**continues access**" in loops. If the access is performed in a noncontinuous way, access time increases due to the organizations of computer architectures. As a result, it makes inefficient computations by noncontinuous accesses.

The order of data storages is different in computer languages. For example, two-dimensional array **A[i][j]** in C language has a direction of continues access in the j-direction. On the other hand, two-dimensional array A(i, j) in Fortran language has a direction of continues access in the i-direction.

For example, in the following code in Fortran:

```
do i=1, n
    A(1, i) = b(i) * c(i)
enddo
```

provides noncontinuous access for the array A.

Let a matrix–matrix multiplication be

$$C = A\,B, \tag{1.2}$$

where the dense matrices are $A, B, C \in \mathfrak{R}^{n \times n}$. Code of matrix–matrix multiplication is as follows:

```
do i=1, n
  do j=1, n
    do k=1, n
      C(i, j) = C(i, j) + A(i, k) * B(k, j)
    enddo
  enddo
enddo
```

In the above program, array **A** is accessed in row direction, which is the second element, k-loop, and array **B** is accessed in column direction, which is the first element, k-loop. Hence, the access of array A is noncontinuous. On the other hand, access of array **B** is noncontinuous in C language.

To perform continuous access for the loop, one method is by making a transportation of array **A**. This is also the same as changing the data structure. This method is effective in many cases. However, one of the drawbacks is the adaptability for actual codes. This is because the actual codes have several thousands of lines, and changing the data structure means several thousands of code modifications. This requires high costs to modify the codes. Hence, changing the data structure of array **A** is impossible in many cases.

In this chapter, we do not consider changing the data structure of the arrays, rather we change the access patterns. In this case, we can replace the order of loops through simple inside computation for the matrix–matrix multiplications. Results are still retained even by replacing the loops to its original form.

With respect to the above discussion, we can consider six kinds of loops, since the loop is three nested. Figure 1.3 shows the six kinds of loops for matrix–matrix multiplication.

In Fig. 1.3, loop structures are categorized in the following forms: (1) **inner product form**, (2) **outer product form**, and (3) **middle product form**. With respect to the continuous accesses in the innermost loop, the middle product form is the best form to obtain high performance.[1]

1.3.2 Data Layout Transformation

As the author explained in the section of continuous loop access, changing the data structure to do continuous access is effective if applied. In this viewpoint, transformation of data layout is one of techniques to speed up.

---

[1]However, the size of problem n is large and data is on out of cache. Hence, the performance will go down even in the middle product form. To avoid this situation, cache blocking, which will be explained later, should be implemented.

Matrix-Matrix Multiplication Code

▸ Inner-product Form
  ▸ Coding with ijk, jik loops.

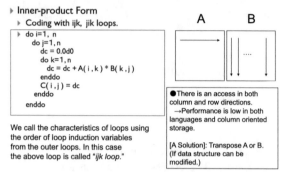

```
▸ do i=1, n
    do j=1, n
      dc = 0.0d0
      do k=1, n
        dc = dc + A( i , k ) * B( k , j )
      enddo
      C( i , j ) = dc
    enddo
  enddo
```

We call the characteristics of loops using the order of loop induction variables from the outer loops. In this case the above loop is called "*ijk loop*."

● There is an access in both column and row directions.
  →Performance is low in both languages and column oriented storage.

[A Solution]: Transpose A or B. (If data structure can be modified.)

(a) Data Access Pattern of Inner-product Form

Matrix-Matrix Multiplication Code

▸ Outer-product form
  ▸ Cording with kij, kji loops.

```
▸ do i=1, n
    do j=1, n
      C( i , j ) = 0.0d0
    enddo
  enddo
  do k=1, n
    do j=1, n
      db = B( k , j )
      do i=1, n
        C( i , j ) = C( i , j ) + A( i , k ) * db
      enddo
    enddo
  enddo
```

● The main access is in the column direction in kji loop.
  → This is good for languages with column oriented storage such as in Fortran.

(b) Data Access Pattern of Outer-product Form

Matrix-Matrix Multiplication Code

▸ Middle-product form
  ▸ Cording with ikj, jki loops.

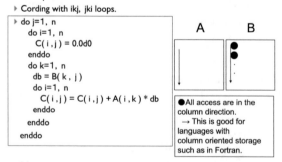

```
▸ do j=1, n
    do i=1, n
      C( i , j ) = 0.0d0
    enddo
    do k=1, n
      db = B( k , j )
      do i=1, n
        C( i , j ) = C( i , j ) + A( i , k ) * db
      enddo
    enddo
  enddo
```

● All access are in the column direction.
  → This is good for languages with column oriented storage such as in Fortran.

(c) Data Access Pattern of Middle-product Form

Fig. 1.3 Category of data access pattern for matrix–matrix multiplications

Let scalars $a, b, c \in \Re$ be the data to be treated. To control these scalars as a data structure, we introduce structure types of arrays for the scalars. The following are known data structures:

- **Array of Structure (AoS)**
  After making a structure of (a, b, c), an array is made with the structure. Let **n** be the number of data elements and **m** be the number of indexes for the array. In this case, **D(3n, m)** is allocated. The first data a is stored in **D(1, 1)**, the second data b is stored in **D(2, 1)**, and the third data c is stored in **D(3, 1)**.
- **Structure of Array (SoA)**
  After making an array **A(m)** for a, an array **B(m)** for b, and an array **C(m)** for c, then a structure is made with the arrays. In this case, **D(n, n, n, m)** is allocated. The first data a is stored in **D(1, *, *, 1)**, the second data b is stored in **D(*, 1, *, 1)**, and the third data c is stored in **D(*, *, 1, 1)**.

With the above explanations, if we access a unit of (a, b, c), which is accessed by (a_1, b_1, c_1), (a_2, b_2, c_2), ..., AoS is performed with a continuous access, while SoA is performed with a noncontinuous access.

On the other hand, if we access only one data from the three elements, such as the access for a in (a, b, c), then this results in a noncontinuous access in AoS.

If the current data structure is extremely unsuitable, and is difficult to modify, you may change the whole programs' data layout just before the computation. After finishing the computation, the layout is returned back to the original structures. Drawbacks of this approach are (1) double amounts of memory are required and (2) the time for data transformation or data copy is required. If the two drawbacks are acceptable in your program, then the approach will be effective.

1.3.3 Organization of Cache and Cache Conflicts

The reason why continuous access provides high performance is because of hierarchical memory. To explain this reason, simple cache organization (a memory model) is explained in this section. Figure 1.4 shows a cache organization.

Figure 1.4 shows that the accessed data in the main memory is automatically moved to the line in cache. This is a function of the hardware, and hence users cannot control this movement.[2] Data is stored as units in the main memory. The unit size is called the **Block Size**. Data movement from the bank of the main memory to the line of cache is controlled by a unit of the block size. There are several methods to put in the data to the lines in cache. We can imagine that there is a mapping for the movement between memory and cache.

The following methods are well used for the mapping.

---

[2]Several computer architectures provide a software control mechanism for the movement of data for cache.

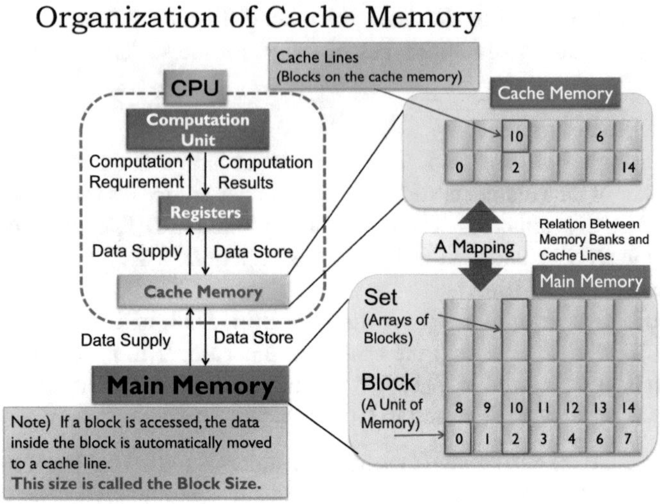

Organization of Cache Memory

Fig. 1.4 A cache organization

For reading from main memory to cache, the following methods are well implemented:

- **Direct Mapping**
  Direct connections from banks in the main memory to the lines in cache are used. This method is allocated with cyclic manner to the lines in cache.
- **Set Associative**
  Flexible allocations between blocks in the main memory and the lines in cache are used. The method where blocks from the main memory can be placed to the lines in cache with arbitrary manner is called "**Full Set Associative**." The method where blocks from the memory can be placed to the lines in cache with N-cycle manner is called "**N-way Set Associative**."

Next, for writing from the lines in the cache to the blocks in the main memory, the following are known:

- **Store Through**
  The method wherein the data is written to a block in the main memory as well as written to a line in the cache.
- **Store In**
  The method wherein the data is written to a block in the main memory while replacement in the target line in the cache is performed.

Let us consider an access that is performed with a noncontinuous access in the main memory with cache organization. Figure 1.5 shows an example of noncontinuous access.

In Fig. 1.5, elements of an array **A(4, 4)** are accessed in a noncontinuous way. The blocks on the main memory are mapped to the lines in the cache by four cycles

Example of Cache Conflict

▸ Let us consider direct mapping
  ▸ Physical connections are as follows
▸ Mapping distance is set to 4.
  ▸ Blocks on main memory are stored to same cache lines with cycle of 4.
▸ Let a line of cache be 8 bytes, and size of blocks be 8 bytes.
▸ Array A is 4x4 with double precision (8 bytes).
   double precision A(4, 4)
  ▸ With this setting, a non-continuous access at a distance of 4 is performed.

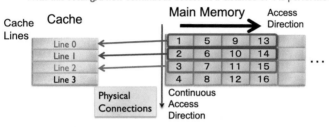

Fig. 1.5   An example of noncontinuous accesses in direct mapping

in direct mapping. Let the number of block sizes in main memory be b and number of lines in cache be c. In the case that four lines are available in the cache, the c is calculated as follows:

$$c = (b - 1) \mod 4. \tag{1.3}$$

Hence, the blocks #1, #5, #9, and #13 are stored to line #0 and the blocks #2, #6, #10, and #15 are stored to line #1, respectively. With this cache organization, we access the first element of A with noncontinuous accesses, where blocks #1, #5, #9, and #13 are accessed. These blocks are stored to the same line #0 due to hardware restriction by direct mapping, and the data on line #0 should be put into the main memory in each access. This is the same as the situation without cache, since we cannot utilize other lines except for the line #0. As a conclusion, the performance is dramatically down. This situation is called as "**Cache Line Conflict**."

The cache line conflict can occur easily in such a case that a number of lines in the cache and dimension of allocated array are the same. The number of lines in cache is organized with a power of two in usual computers. Hence, the cache line conflict may occur if the size of the allocated array is in the power of two, such as 1024.

A "**patting**" is known as a technique to avoid cache line conflict. This allocates an additional space to avoid a size to the power of two. Then, computations are performed to the allocated array in the same manner. The patting is also available for compiler options. Hence, another solution is to use the compiler option which can apply the patting for code optimization.

1.3.4 Loop Unrolling

Loop unrolling is a technique to put data into registers, or to use high-efficient instructions, such as SIMD operations, through code optimizations by a compiler.[3]

Let us consider a code of matrix–matrix multiplications. In this code, the loop unrolling is used to unroll the loop with stridden more than one. Let the stride of loops be *m*. In this case, loop unrolling is applied to the loop with depth of *m* or **m-th loop unrolling**. The non-unrolling code is as follows:

```
do i= 1 , n
  do j= 1 , n
    do k= 1 , n
      C(i, j) = C(i, j) +A(i, k) *B(k, j)
    enddo
  enddo
enddo
```

The loop unrolled code with depth of two for the k-loop is as follows:

```
do i= 1 , n
  do j= 1 , n
    do k= 1 , n, 2
      C(i, j) = C(i, j) +A(i, k) *B(k, j) + A(i, k+1)*B(k+1, j)
    enddo
  enddo
enddo
```

By adapting loop unrolling with a depth of two, the number of branch judgements for k-loop is reduced from n to n/2, thus execution time may be reduced.[4]

Please note that the case above is when *n* can be divided by 2. If *n* cannot be divided by 2, programmers need to consider computations for the remainder part. In detail, programmers need to write the codes as follows, considering loop unrolling with a depth *m*.

---

[3] Another merit of loop unrolling is to reduce the time of branch control in loops.

[4] However, the efficiency of loop unrolling depends on code optimizations by the compiler. To do an effective implementation of loop unrolling, programmers need to know in advance the details of the code optimization by the compiler.

```
! < Loop Unrolling with Depth of m >
do k=1, n, m
   ...
enddo
! < A Reminder Loop for n / m. >
if (mod(m,n) .neq, 0) then
   do k = (n/m)*m, n
      ...
         < Original code of computation part >
   enddo
endif
```

Next, code of loop unrolling with a depth of two for j-loop is as follows. The following code is where n can be divided by 2:

```
do i= 1 , n
   do j= 1 , n, 2
      do k= 1 , n
         C(i, j ) = C(i, j ) +A(i, k) * B(k, j )
         C(i, j+1) = C(i, j+1) +A(i, k) * B(k, j+1)
      enddo
   enddo
enddo
```

To make the same result as with the original code, the two statements are added in the innermost loop as shown above. The common elements of **A(i, k)** appear in the statements. **A(i, k)** is stored in the main memory if the code optimization is not applied, but it can also be stored in a register through code optimization. With this register allocation, access time for **A(i, k)** is dramatically reduced the next time you access it.

The above code gives us a nice example to speed up by applying loop unrolling through code optimization in the compiler.

On the other hand, loop unrolling can also be applied for the outermost loop of k-loop in the matrix–matrix multiplication code. The code is shown as follows:

```
do i= 1 , n, 2
   do j= 1 , n
      do k= 1 , n
         C(i , j) = C(i , j) + A(i , k) * B(k, j)
         C(i+1, j) = C(i+1, j) + A(i+1, k) * B(k, j)
      enddo
   enddo
enddo
```

The readers can understand that the above code can also be optimized by storing **B(k, j)** to a register.

Interestingly, loop unrolling can be applied simultaneously for two loops in the matrix–matrix multiplications. The following is the loop unrolling code with depth of two for i-loop and j-loop:

```
do i= 1 , n, 2
  do j= 1 , n, 2
    do k= 1 , n
      C(i ,j ) = C(i , j ) + A(i , k) * B(k, j )
      C(i ,j+1) = C(i , j+1) + A(i , k) * B(k, j+1)
      C(i+1,j ) = C(i+1, j ) + A(i+1, k) * B(k, j )
      C(i+1,j+1) = C(i+1, j+1) + A(i+1, k) * B(k, j+1)
    enddo
  enddo
enddo
```

The above code indicates that the four elements of **A(i, k)**, **A(i + 1, k)**, **B(k, j)**, and **B(k, j + 1)** can be stored in registers.

As explained above, multiple data can be stored in registers by applying loop unrolling for appropriate loops. Therefore, we can reduce the time of data movements from the main memory. The optimal number of variables should be stored in registers which depends on the number of registers in the computer architecture.

If the number of variables to be stored in registers is larger than the actual number of registers, to adapt unrolling too many times, data is stored back to the main memory. This situation is called "**register spill**." If register spill occurs, the performance goes down.

1.3.5 Cache Blocking (Tiling)

Mainly, loop unrolling is a technique to optimize data movement to registers. Hence, if the areas of access are wide, the execution time slows down due to a miss of data from cache lines even in a case where loop unrolling is adapted. In this section, we explain a solution for this data miss on cache.

Even if the data access pattern is modified to do continuous access, the data miss from the cache will occur when the area of access is as large as the cache lines. The miss of data on lines of cache is called "**Cache Miss–hit**."

To reduce cache miss–hits, loop length should be shortened and the loop should be divided into multiple loops and reduce access region of computations. It depends on the sentences in the innermost loop whether this modification should be adapted or not. Matrix–matrix multiplication can adapt this modification because sentences for the innermost loop are simple.

Modifying programs or algorithms to reuse data in lines on cache are called "**Cache Blocking**." Dividing loop into small pieces of loops to establish cache blocking is called "**Tiling**." These are important techniques to obtain high performance on recent computers.

Let us consider applying the blocking to a code of matrix–matrix multiplications. Let **ibl** be the size of blocking. Since the statement inside the loop is simple for matrix–matrix multiplications, we can adapt the blocking for all three-nested loops. The following is the blocking code:

```
do ib=1, n, ibl
  do jb=1, n, ibl
    do kb=1, n, ibl
      do i=ib, ib+ibl-1
        do j=jb, jb+ibl-1
          do k=kb, kb+ibl-1
            C(i, j) = C(i, j) + A(i, k) * B(k, j)
          enddo
        enddo
      enddo
    enddo
  enddo
enddo
```

Multiple small matrix–matrix multiplications with **ibl** × **ibl** are performed to compute whole matrix–matrix multiplications. Since whole matrix–matrix multiplications are divided by multiple small matrix–matrix multiplications, the number of cache miss–hits can be reduced. Thus, blocking is established by the above loop. Figure 1.6 shows a snapshot of the blocking.

The best size of **ibl** in Fig. 1.6 depends on computer architectures where the total amount of lines on cache affects the best size of **ibl**. The best **ibl** should be determined with respect to the size of small matrices for **A**, **B**, and **C**.[5]

Since the blocking of cache and loop unrolling are different concepts, both are simultaneously adaptable for the target loop. Loop unroll can be adapted for the

---

[5]In some numerical libraries, the best blocking size to do matrix–matrix multiplications is fixed by library developers in advance before the release time. Hence, library users do not tune the blocking size. In addition, some numerical libraries perform test execution to tune the blocking size automatically with respect to the amount of cache and considering computations of target loops in install time. These automations are called "**Software Auto-tuning**." [3–7] Study on software auto-tuning is one of hot topics in high-performance computing. In Chap. 5, we will explain the details of the research on software auto-tuning.

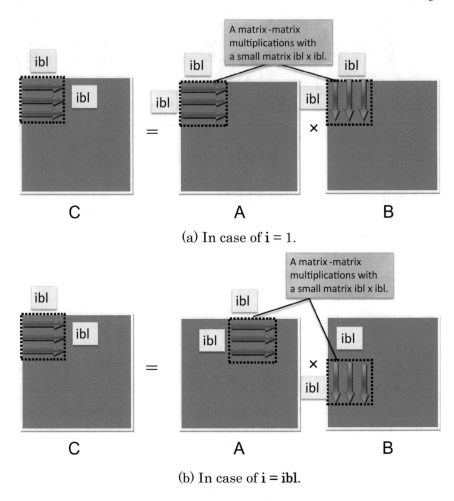

(a) In case of **i** = 1.

(b) In case of **i** = **ibl**.

Fig. 1.6 A snapshot of blocking for matrix–matrix multiplications

six-nested loop for blocking. Adapting for both cache blocking and loop unrolling is established and the data are stored in registers and data to write are stored in lines on cache. Hence, extremely high performance will be obtained.

The following code is an example for adaptation of blocking and unrolling between i-loop and j-loop with depth of two:

```
do ib=1, n, ibl
   do jb=1, n, ibl
      do kb=1, n, ibl
         do i=ib, ib+ibl, 2
            do j=jb, jb+ibl, 2
               do k=kb, kb+ibl
                  C(i , j) = C(i , j) + A(i , k) * B(k, j)
                  C(i+1, j) = C(i+1, j) + A(i+1, k) * B(k, j)
                  C(i ,j+1) = C(i , j+1) + A(i , k) * B(k, j+1)
                  C(i+1,j+1) = C(i+1, j+1) + A(i+1, k) * B(k, j+1)
               enddo
            enddo
         enddo
      enddo
   enddo
enddo
```

1.3.6 Loop Transformations

The change of loop structure to optimize register allocation of data or increase parallelism is called "**Loop Transformations.**" In this section, some major optimization methods in loop transformations are explained.

We have the following code as follows:

```
DO K = 1, NZ
   DO J = 1, NY
      DO I = 1, NX
         RMAXY = 4.0/(1.0/R(I,J,K)+1.0/R(I+1,J,K)+1.0/R(I,J+1,K)+1.0/R(I+1,J+1,K))
         RMAXZ = 4.0/(1.0/R(I,J,K)+1.0/R(I+1,J,K)+1.0/R(I,J,K+1)+1.0/R(I+1,J,K+1))
         QG = ABSX(I)*ABSY(J)*ABSZ(K)*Q(I,J,K) SXY(I,J,K) =
              (SXY(I,J,K)+(RMAXY*(DXVY(I,J,K) +DYVX(I,J,K)))*DT)*QG
         SXZ(I,J,K) = (SXZ(I,J,K)+(RMAXZ*(DXVZ(I,J,K) +DZVX(I,J,K)))*DT)*QG
      ENDDO
   ENDDO
ENDDO
```

Loop Fusion or Loop Collapse

The first example of loop transformation is **Loop Fusion** or **Loop Collapse**. Loop fusion is a technique to fuse multiple loops to make it as one loop or reduced loops. Loop collapse is a technique to break down loop structures. It is a technique to transform a long-length loop to its original length to reduce time of reading data from the main memory, or to obtain higher parallelism when the loop is captured with parallelization.

The following code shows a loop collapse to the code shown above:

```
DO KK = 1, NZ * NY * NX
    K = (KK-1)/(NY*NX) + 1
    J = mod((KK-1)/NX,NY) + 1
    I = mod(KK-1,NX) + 1
    < Original Computations >
ENDDO
```

In the code shown above, three-nested loop in original code is collapsed to make a one-nested loop. The original loop has only NZ for the K-loop, while the collapsed loop has NZ * NY * NX for KK-loop. This extension affects parallel processing for the outer loops, such as thread parallelization by OpenMP. This will be discussed in Chap. 2.

One of the drawbacks for the above collapsed loop is that the compiler does not understand the continuous nature of KK-loop, since indexes of K, J, and I are calculated by KK. For this reason, the compiler may not generate optimized code for the loop. To solve this problem, the following loop collapse may be effective in some computer environments:

```
DO KK = 1, NZ * NY
    K = (KK- 1 )/NY + 1
    J = mod(KK- 1 ,NY) + 1
    DO I = 1, NX
        < Original Computations>
    ENDDO
ENDDO
```

For the above loop collapsed code, the length of the outer loop is increased and the innermost loop is continuous. Hence, it does not hinder compiler optimizations, where the above collapsed two-nested loop may be better than the collapsed one-nested loop.

Loop Split

Below shows an example of loop split. Application of loop split to the original code is as follows:

```
DO K = 1, NZ
  DO J = 1, NY
    DO I = 1, NX
      RMAXY = 4.0/(1.0/R(I,J,K)+1.0/R(I+1,J,K) +1.0/R(I,J+1,K)+1.0/R(I+1,J+1,K))
      QG = ABSX(I)*ABSY(J)*ABSZ(K)*Q(I,J,K)
      SXY(I,J,K) = (SXY(I,J,K)+(RMAXY*(DXVY(I,J,K) +DYVX(I,J,K)))*DT)*QG
    ENDDO
    DO I = 1, NX
      RMAXZ = 4.0/(1.0/R(I,J,K)+1.0/R(I+1,J,K)+1.0/R(I,J,K+1)+1.0/R(I+1,J,K+1))
      QG = ABSX(I)*ABSY(J)*ABSZ(K)*Q(I,J,K)
      SXZ(I,J,K) = (SXZ(I,J,K)+(RMAXZ*(DXVZ(I,J,K)
                +DZVX(I,J,K)))*DT)*QG
    ENDDO
  ENDDO
ENDDO
```

The computations in the loop body are reduced by splitting I-loop in the above code. If there are register spills in the original code due to many computations in the loop body, the above split code provides better performance through register optimizations to avoid register spills. On the other hand, the above code needs to recompute QG, since lack of QG happens in the second loop by the loop split. This causes additional overhead, which means an increase in computation complexity and data access to that of the original code. If the performance including the overhead for the split code is acceptable, the split code can be used to the original code.

On the other hand, the loop split shown above can apply to the K-loop. This splitting makes a perfect split forming two-separate and three-nested loops. This is another way to do loop split for the original code.

1.3.7   Another Technique to Speed up

In this section, general techniques to speed up are explained.

Reduction of Common Parts
The following statements have a room to optimize performance in which there are common elements in each statement:

$$d = a + b + c$$
$$f = d + a + b$$

With respect to "**a + b**", the following code for speedup can be derived:

$$\text{temp} = a + b;$$
$$d = \text{temp} + c;$$
$$f = d + \text{temp};$$

Let us consider the following:

do i=1, n

  xold(i) = x(i)

  x(i) = x(i) + y(i)

enddo

The array **x(i)** above is regarded as a common part. Hence, the following code can be derived:

do i=1, n

  dtemp = x(i)

  xold(i) = dtemp

  x(i) = dtemp + y(i)

enddo

Movement of codes

Let us consider the following code:

do i=1, n

  a(i) = a(i) / dsqrt(dnorm)

enddo

Generally speaking, the execution time for operating divisions is much slower than execution time for multiplications. Hence, the above code is extremely slow if the division **dsqrt(dnorm)** is performed in each loop iteration. If programmers guarantee accuracy of results in computations, the following change of code can be acceptable:

dtemp = 1.0d0 / dsqrt(dnorm)

do i=1, n

  a(i) = a(i) * dtemp

enddo

For the above code, the division is changed to a multiplication with "**a(i) * dtemp**." Hence, it has a potential to speed up.

Removal of IF-sentences Inside Loops:

If there are "IF-sentences" in loops, particularly in innermost loops, code optimizations of compilers or hardware are prevented. For example, a branch predictor in hardware is constructed with "taken" for IF-sentence to speed up, but the actual judgement of the branch is "not-taken", then the execution slows down. Some compilers provide effective codes by applying pipelining of data from the main memory to the registers provided that there are no IF-sentences; otherwise, pipelining is not applied. Hence, it is better to remove IF-sentences inside the loop as possible as you can.

In removing IF-sentences, it is easier to move the IF-sentences inside the loop if it is unnecessary. However, algorithm modification is required in many cases if the IF-sentences are essentially needed for the algorithm.

Here, the following code is set to "1" for the diagonal elements of array **A**; otherwise, they are set to elements of array **B**:

```
do i=1, n
  do j=1, n
    if ( i .neq. j ) then
      A(j, i) = B(j, i)
    else
      A(j, i) = 1.0d0
    endif
  enddo
enddo
```

An IF-sentence performed in every iteration of loops is shown above.

To remove the IF-sentence, the algorithm should be modified as follows: (1) Making a loop for setting elements of array **B** with elements of array **A** and (2) making a loop for setting diagonal elements of array **A** with 1. The code is shown as follows:

```
do i=1, n
  do j=1, n
    A(j, i) = B(j, i)
  enddo
enddo
do i=1, n
  A(i,i) = 1.0d0
enddo
```

With the above code, speedup will be attained in several computer environments.

Strengthen Software Pipelining:
As mentioned, there is a software pipelining to reduce time for data movement from the main memory to the registers. An example for a code of loop unrolling with a depth of two is shown below:

```
do i=1, n, 2
   dtmp_b0 = b(i)
   dtmp_c0 = c(i)
   dtmp_a0 = dtmp_b0 + dtmp_c0
   a(i) = dtmp_a0
   !--- code of loop unrolling with depth of two.
   dtmp_b1 = b(i+1)
   dtmp_c1 = c(i+1)
   dtmp_a1 = dtmp_b1 + dtmp_c1
   a(i+1) = dtmp_a1
enddo
```

The above code is not optimized for sentences in loops but it can be done by applying loop unrolling. For example, after reading **dtmp_b0** and **dtmp_c0**, the computation of "**dtmp_b0 + dtmp_c0**" is performed by just reading the data. Hence, it is not capable to optimize the code for both reading and using. To apply code optimization, modification of the order of sentences is applied as follows:

```
do i=1, n, 2
   !--- Reading of b().
   dtmp_b0 = b(i)
   dtmp_b1 = b(i+1)
   !--- Reading of c().
   dtmp_c0 = c(i)
   dtmp_c1 = c(i+1)
   !--- Computation Parts.
   dtmp_a0 = dtmp_b0 + dtmp_c0
   dtmp_a1 = dtmp_b1 + dtmp_c1
   a(i) = dtmp_a0
   a(i+1) = dtmp_a1
enddo
```

In the code above, distance between the reading of **b()** and the **c()** to using them is larger than the distance of the original. As a result, it still has a room to apply pipelining in the code. Hence, the above code has the potential to speed up.

1.4   Use of Numerical Library

If the program is using the same function to numerical library, the user should utilize the function of the library provided by specialists or vendors, rather than his/her own codes. In particular, computations of dense matrices have a much room to optimize, thus users should use the function of a library.

Basic Linear Algebra Subprograms (BLAS) is a standard Application Programming Interface (API) for dense linear algebra operations. Computations are categorized as follows:

- Level 1 BLAS: A Category of dot products, vector and scalar operations, etc.
- Level 2 BLAS: A Category of matrix–vector multiplications, etc., and
- Level 3 BLAS: A Category of matrix–matrix multiplications, etc.

The Level 3 BLAS has much room to optimize codes, since it uses matrix–matrix operations. Computer complexity of the matrix–matrix operations is $O(n^3)$ and the memory space is $O(n^2)$, and hence data reusability is very high. Reusability is the reason why cache blocking can be used for matrix–matrix multiplication.

DGEMM is one of the Level 3 BLAS operations. It is described as the following function:

$$C := \text{alpha} * \text{op}(A) * \text{op}(B) + \text{beta} * C,$$

where the sizes of arrays A, B, and C in the above are A: M * K, B: K * N, and C: M * N, respectively. Interface of DGEMM in the viewpoint of BLAS library is as follows:

CALL DGEMM("N", "N", n, n, n, ALPHA, A, N, B, N, BETA, C, N).

Description of the elements from the left-hand side is given in the following: A is transpose or not, B is transpose or not, size of M, size of N, size of K, value of alpha, address of A, number of elements for the first dimension of A, address of B, number of elements for the first dimension of B, value of beta, address of C, and number of elements for the first dimension of C, respectively. The interface is difficult to understand for nonexpert users for BLAS, but using it gives us high performance. The author will explain performance trends of typical performance of BLAS. Figure 1.7 shows the trends of performance for BLAS.

Figure 1.7 indicates that Level 3 BLAS (BLAS3) reaches almost peak performance when the size of the matrix is large. The peak performance of BLAS3 is more than 90% of the peak experimentally. On the other hand, BLAS1 and BLAS2 have low efficiencies to the peak. Due to the cache amount, the performance of BLAS1 and BLAS2 affect the usual amounts of hierarchical caches, such as the size of L1 cache and LLC.

The performance of BLAS in the case of a small-sized matrix should be taken into consideration. Many implementations of BLAS focus on the performance in large-sized matrix because it is difficult to obtain high performance in a small-sized matrix. Small size refers to a size of less than 100×100 experimentally. Take note

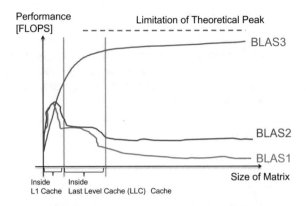

Fig. 1.7 Trends of typical performance of BLAS

that when we parallelize the matrix with thread execution, the size of the matrix is less than 100×100 per thread even if we execute several thousands of matrices in the viewpoint of a whole size.

LAPACK, which can solve linear equations and eigenvalue problems for dense matrices, and ScaLAPACK, which is a dense numerical library on distributed memory machines are very useful tools and can easily establish high performance for nonexpert users. The author recommends using such libraries for users to operate dense matrices.

Exercises

1. Survey performance between continuous accesses and noncontinuous accesses for matrix–matrix multiplications with a computer.
2. Evaluate performance of loop unrolling for i-loop, j-loop, and k-loops for matrix–matrix multiplications.
3. Survey performance of loop unrolling with a blocking code for matrix–matrix multiplications.
4. Check the number of registers on the computer nearby, the organization of the cache, and the size of lines on the caches. Consider the performance for the above 1–3 with respect to these components of hardware.

References

1. X.S. Li, J.W. Demmel, D.H. Bailey, G. Henry, Y. Hida, J. Iskandar, W. Kahn, S.Y. Kang, A. Kapur, M.C. Martin, B.J. Thompson, T. Tung, D.J. Yoo. ACM Trans. Math. Softw. **28**, 152 (2002)
2. A. Buttari, J. Dongarra, J. Kuzak, P. Luszczek, S. Tomov. ACM Trans. Math. Softw. **34** (2008)
3. T. Katagiri, K. Kise, H. Honda, Yuba, in *Proceedings of the ACM International Symposium on High Performance Computing* (2003), pp. 146–159

4. T. Katagiri, K. Kise, H. Honda, T. Yuba, in *Proceedings of the ACM Computing Frontiers* (2004), pp. 12–25
5. T. Katagiri, K. Kise, H. Honda, T. Yuba, Parallel Comput. **32**(1), 92–112 (2006)
6. K. Naono, K. Teranishi, J. Cavazos, R. Suda, *Software Automatic Tuning* (Springer, New York, 2010)
7. T. Katagiri, S. Ohshima, M. Matsumoto, in *Proceedings of IEEE MCSoC 2014* (2014), pp. 91–98

Chapter 2
Basics of MPI Programming

Takahiro Katagiri

Abstract This chapter explains the basics of parallel programming and MPI (Message Passing Interface). To understand MPI, parallel computer architectures and parallel programing models are explained in the first chapter. APIs (Application Programming Interfaces) of MPI are shown with examples of MPI programming. Several key topics, such as data distribution methods and communication algorithms for MPI programming are also explained.

2.1 Parallel Programming

2.1.1 Why Do We Parallelize Programs?

The main reason why users want to parallelize their program is reducing execution time for the program.[1] Let the sequential execution time for a program be T. The aim of parallelization is to establish:

$$\text{Execution time is reduced to } T/p \text{ with } p \text{ computers.}$$

In theory, it is easily done, but in practice, the execution time depends on the nature of the program/algorithm. Hence the difficulty of parallelization is different for each situation to be parallelized. One of the main reasons for this is the sequential part inside the program, which also includes communication. There is a setup time for the communication, and time for the transfer of data in the communication. Due to this reason, it is difficult to parallelize a code in general.[2]

---

[1] On the other hand, using large memory is another reason to parallelize a program by using distributed memory machines.

[2] But the communication can be parallelized with respect to algorithm. Explanation of this kind of parallelization will be done in this chapter.

T. Katagiri (✉)
Information Technology Center, Nagoya University, Nagoya, Japan
e-mail: katagiri@cc.nagoya-u.ac.jp

© Springer Nature Singapore Pte Ltd. 2019
M. Geshi (ed.), *The Art of High Performance Computing for Computational Science, Vol. 1*, https://doi.org/10.1007/978-981-13-6194-4_2

In this chapter, we summarize the essential concepts of parallel programming by understanding the different kinds of parallel machines and models of parallel programming

2.1.2 Classification of Parallel Computers

According to a classification by Professor Michael J. Flynn at Stanford University in 1966, parallel computers are classified by the following categories:

- SISD (Single Instruction Single Data Stream)
- SIMD (Single Instruction Multiple Data Stream)
- MISD (Multiple Instruction Single Data Stream)
- MIMD (Multiple Instruction Multiple Data Stream).

In the above classification, parallel computers are classified as single or multiple instructions, and single or multiple data streams. All of the current parallel computers are classified as MIMD. However it is very difficult to understand the MIMD in the viewpoint of parallel programming. Because of this, the more commonly used model for parallel programing is SIMD, which is also used for instruction level parallelism. Hence the readers are expected to be familiar with the concept of SIMD.

On the other hand, there is a different classification set for hardware organization. Figure 2.1 shows the classification for memory organizations.

In Fig. 2.1a, a computational element is denoted by the term PE (Processing Element), which is usually denoted as the "core" in the CPU. The Shared Memory Type provides a large memory (a shared memory), and each PE can access the shared memory. In particular, a parallel computer that can access each PE with the same execution time is called "SMP (Symmetric Multiprocessor)." The memory organization in SMP is called "UMA (Uniform Memory Access)." It is said that programming with shared memory type is easy since a programmer can use a shared array on a program. This comes from the memory that is not distributed. On the other hand, hardware cost for shared memory type is very high, because of the complex organization (circuits) of the hardware. Limited parallelism for shared memory type can be provided even up to now, such as up to a thousand of parallelism or so.

The parallel computers in the Distributed Memory Type in Fig. 2.1a consist of a distributed memory. There is no way to access this distributed memory directly in the hardware. To know the contents of the distributed memory, a message passing to PE with the target distributed memory is performed. Due to this process, parallel programming becomes more difficult. However, parallel computers with distributed memory can consist of a million parallelisms. All parallel machines with a high number of parallelisms are the distributed memory type; hence we should take note of the programming done on distributed memory type.

The parallel computers of DSM (Distributed Shared Memory) Type in Fig. 2.1a provide software or hardware mechanism to establish shared memory with a physically distributed memory. The DSM type shares merit between the shared memory

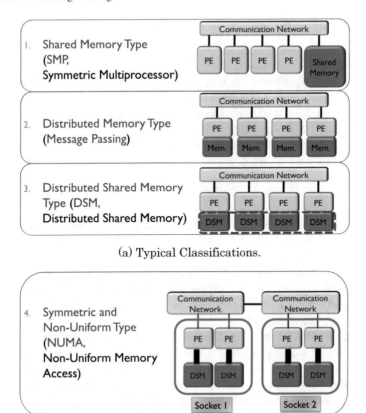

(a) Typical Classifications.

(b) NUMA Organization.

Fig. 2.1 Classification of parallel computers for memory organizations

and the distributed memory types. However, the speed of access for the distributed memory is extremely slow compared to that of the shared memory type. Hence parallel computing performance is poor in some cases compared to the shared memory type.

Recently, many parallel computers with NUMA (Non-Uniform Memory Access) in Fig. 2.1b are becoming more common. This is due to the high usage of hardware resources and high efficiency of power for NUMA machines compared to UMA machines. In a NUMA machine, access to near memory is faster than access to far memory for each PE. PEs which can access the memory with the same access times are called "PEs on **socket**." Recently, NUMA machine that consists of two to four sockets are widely-used, and the number of sockets tends to be increased.

2.1.3 Model of Parallel Programming

As explained previously, the actual behavior of parallel programs is MIMD, which is difficult to use as a programming model due to its complexity. Hence SIMD is used as a programming model for usual programming.

SIMD is a model providing similar computations, such as addition, with multiple input data, but it outputs multiple data. For example, we have data of (1, 2, 3, 4) for input. Then SIMD computes multiple additions of (1 + 2, 3 + 4) simultaneously. After the addition (3, 7) are outputted simultaneously. Considering that the input data is in the form of an array, the instruction is in the form of a loop, and the output data is in the form of an array, SIMD is also providing a program model to make a parallel programming.

SPMD (Single Program Multiple Data) is a programming model that uses a single program to execute parallel processing. SPMD has the same concept to SIMD and is used for a programming model for MPI version 1. Hence SIMD can apply the concept to the previous example, which is a relationship between arrays and loops. Since the target program is only one program, common computations and arrays are performed in the same loop. To do parallel processing, different parts of the allocated arrays are accessed in each PE.

The readers need to understand the differences between processes and threads when used in a parallel processing manner. The process is a concept of different memory space, and the thread is a concept of shared memory space. Hence execution by MPI is categorized as a process. On the other hand, execution by OpenMP, which will be explained in Chap. 3, is categorized as a thread.

Since process and thread are different concepts, we can use a hybrid execution on both. This hybrid execution is performed by process execution with different memory, and then by thread execution with shared memory in the execution process. This hybrid execution is called "hybrid MPI/OpenMP execution," if we use MPI and OpenMP. The hybrid MPI/OpenMP execution will be explained in a later chapter.

2.1.4 Metrics of Performance Evaluation

To do performance evaluation in parallel programming, how should we evaluate it? The answer is "speedup" by parallel computing. Let Ts be the time for sequential execution, and Tp be the time for parallel programming with P, which is the number of parallel processing. We can define the speedup as:

$$Sp = Ts/Tp \ (0 \leq Sp), \tag{2.1}$$

where $Sp = P$ with P parallel processing, we call this an "ideal speedup." If we obtain $Sp > P$ with P parallel processing, we call this a "super linear speedup." The main reason that results in a super linear speedup is cache effect—the access area

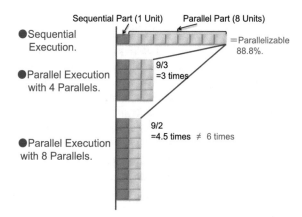

Fig. 2.2 An example of Amdahl's law

is shortened by parallel processing, then the data to be used is on cache. Hence as a result, we obtain a higher computational efficiency to the original.

If we need to calculate the efficiency of parallel processing, the following equation can be used:

$$Ep = Sp/P * 100\,(0 \le Ep), \tag{2.2}$$

where the unit is [%]. By using the formula in Eq. (2.2), we can evaluate solid difference to ideal efficiency of parallelism.

The speedup factor has a property of saturation. The saturation value is the upper limit of speedups when we increase the parallelism of P.

There is a well-known theorem for parallel processing, named Amdahl's law. Let the time of sequential processing be K, and the ratio of parallel parts in sequential processing be α. With these parameters, the speedup can be computed by:

$$S_P = \frac{K}{\frac{K\alpha}{P} + K(1 - \alpha)} = \frac{1}{\frac{\alpha}{P} + (1 - \alpha)} = \frac{1}{\alpha\left(\frac{1}{P} - 1\right) + 1} \tag{2.3}$$

Equation (2.3) shows that the upper limit of speedup is $1/(1 - \alpha)$ if we use an infinite number of parallelisms ($P \to \infty$). This implies that if we parallelize a program with 90% parallelism inside the program, the limit of parallelism in this program is $1/(1 - 0.9) = 10$. The number of speedup does not change even if we use 100 computers. Figure 2.2 shows this situation with another example.

Figure 2.2 shows that three times speedup is obtained when the ratio of parallel parts is 88.8% in 4 parallels. This is a good ratio for parallelism. However, we cannot say that 3 times $\times$ 2 = 6 times speedup in the case of 8 parallels, but only 4.5 times is obtained in this case. This indicates that the ratio of the sequential part is getting larger as the number of parallels increases.

To measure speedups, the following two metrics are known.

- Strong Scaling: After the size of problems in global view is fixed, parallelism of p increases. The explanation in this chapter of parallel processing is based on the strong scaling. The increase of communication time (since the number of parallelism p increases) means it is difficult to obtain high efficiency of speedup.
- Weak Scaling: After the size of problems in unit of parallelisms, such as per node, is fixed, parallelism of p increases. In weak scaling, computational complexity is a constant for some algorithms. Hence the execution time in weak scaling is expected to a reach constant time even when the number of parallelisms increases, such as when a target program has $O(N)$ computational complexity in local view. On the other hand, if the target program has $O(N^3)$ computational complexity in global view, it is impossible to keep constant time due to the increase of the number of parallels. Since the total size of problems is increasing in this situation, the total computational complexity increases. As a conclusion, weak scaling cannot be adaptable for this case.

2.1.5 Data Distribution Methods

One of the important concepts for parallelizing a program is data distribution. This is the same as data structure in sequential processing. The readers may have learned that the data structure is one of the important concepts for sequential processing, but this is also the case for parallel processing. One of the reasons that data distribution is important is **load balancing**—it affects the total efficiency of parallelism. The data is allocated by arrays in the program, hence how to distribute the arrays to distributed memory, or how to owe computations for part of the allocated arrays to cores with shared memory, are some of the major issues in parallel processing. An example of data distribution is shown in Fig. 2.3.

In Fig. 2.3, there are two ways of data distribution manners, such as block and cyclic. In addition, there are also two ways for distribution directions, such as 1D and 2D.

For the 1D distribution in Fig. 2.3a, there are three types of distribution methods as follows:

- **Block Distribution**: The lines of the array are distributed with constant lines. Let N be the number of rows, and the number of PEs is 4. In this case, the number of each line is set to N/4. If N/4 is not divisible, the remainder lines go to a PE. Usually, the last PE, such as PE #3, collects the remainder in this case.
- **Cyclic Distribution**: The number of allocated lines is set to 1. If the line #4 is allocated to PE #3, then PE #0 is allocated line #5. In this way, the allocation is performed in a cyclic manner.
- **Block-Cyclic Distribution**: The number of allocated lines is set to a block length, such as 2, but is allocated in a cyclic manner. This provides intermediate characteristics between block and cyclic distributions.

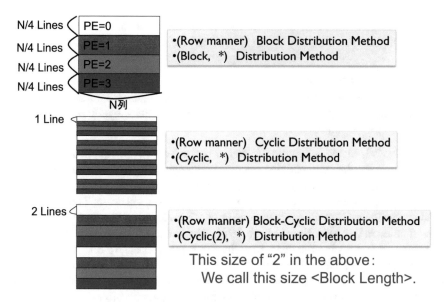

(a) 1D Distribution.

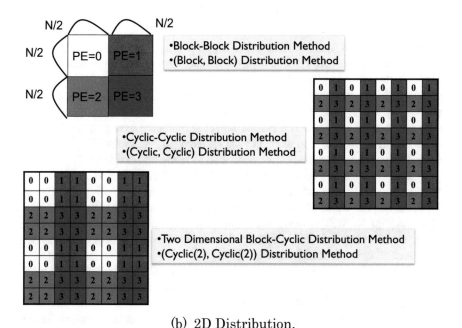

(b) 2D Distribution.

Fig. 2.3 Examples of data distribution methods

On the other hand, Fig. 2.3b shows 2D distribution with three kinds of distributions. They operate similarly just as they would for a 1D case. The followings are known.

- **Block-Block Distribution**: The block distribution is adapted for 2D directions.
- **Cyclic-Cyclic Distribution**: The cyclic distribution is adapted for 2D directions.
- **Two Dimensional Block-Cyclic Distribution**: The block-cyclic distribution is adapted for 2D directions.

In Fig. 2.3, block distribution seems to be better for many applications, but it depends on the data computation pattern. If the data access is unbalanced for the array, such as ones with very heavy computations that are performed with the right-down part, and we use block distribution, load imbalance will occur. This is because PE #3 has the most of all computations in Fig. 2.3a. Hence it is better to use cyclic distribution in this case. However, cyclic distribution causes heavy communications in some programs since the distributed line has to cycle through a line, and the other PEs might want to obtain data from each distributed line. In such a case, block-cyclic distribution is the better option. In this case, the **block length** becomes an important parameter of performance to minimize the time of communications.

2.1.6 Characteristics of MPI

MPI [1, 2] is a standard interface for message passing, and it is also a model of message passing. MPI is not an interface of compilers, dedicated software or libraries, but it provides the functions of an interface.

MPI is designed for parallel processing on distributed memory parallel machines, and is good for large-scale computations. By using parallelization with MPI, it can overcome limitation of memory size on a node or limitation of treatments for large-scale file sizes. MPI is also available for massive parallel processing, and it can dramatically shorten execution time compared to sequential execution.

One of the most attractive merits is portability. API (Application Programming Interface) for MPI is standardized. Hence, when the readers develop a code with MPI, the readers can execute the code in several parallel environments that have been installed for MPI, without modification. All supercomputers and PC clusters are installed or can be installed with MPI, and therefore, the benefit of having this high portability is a critical feature of MPI.

From the point of view of performance, it can establish high scalability of parallelism if the algorithm for the code has high parallelism. This scalability means that high-performance can be obtained by using parallelisms.

MPI can optimize codes for communications with direct descriptions of API by users. However, the effect of optimizations depends on the skills of the user. The cost of programming with MPI increases with respect to sequential programming, but the author thinks this is caused by a lack of experience using MPI. Hence if the user can obtain some basic skills of MPI, the cost does not increase so much compared to

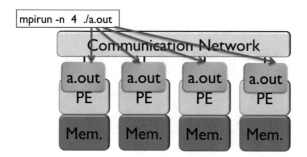

Fig. 2.4 A MPI execution

the others. The author thinks that there is no cost difference between MPI and other approaches if the user wants to obtain high-performance for parallel processing.

In this textbook, we explain MPI version 1 (hereafter, we describe MPI-1). Several extensions are made in MPI version 2, such as execution on a master-worker model. Additional extension of functions is also made in MPI version 3, such as non-blocking communications for dedicated MPI functions.

As the author mentioned before, MPI-1 is based on the SPMD model. Hence MPI-1 starts its execution by copying one program to all processes of MPI in run-time. Here, we have an executable code, named "a.out." Execution of MPI is shown in Fig. 2.4.

In Fig. 2.4, a command mpirun is used for starting MPI. To specify number of parallelisms, option "-n" should be specified. There is another way to specify the number of parallelisms in supercomputer environment. In this case, option "-n" can be omitted.

In Fig. 2.4, the readers can see that different memory spaces are allocated in each process for an array, which is using the same name of the array in the program after starting **a.out**.

2.2   Basic MPI Functions

2.2.1   Explanation of MPI Terminology

MPI communicates among processes. Each process is allocated to each core as 1-to-1 allocation between them if we do not use simultaneous processing technologies, such as HT (Hyper Threading).

"Rank" is one of the most important terminologies in MPI. Rank is an identifier number of process in each MPI process. Number of rank is stored in a variable which is specified in **MPI_Comm_rank** function. It starts from 0 to (number of parallelisms)−1, where the number of parallelisms is specified by user in start-time.

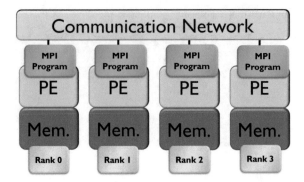

Fig. 2.5 An explanation of Rank

Figure 2.4 shows an explanation of rank.

In Fig. 2.5, each PE is allocated with different numbers.

To know the number of processes in the world of MPI, function **MPI_Comm_size** is used.

"blocking" function in the terminology of MPI means a receiving function does not return to program after calling the corresponding sending function until receiving data, and the data is fully copied to a receive buffer. To understand the **blocking function** in MPI, **communication modes** should be considered. During execution with a default communication mode, **standard communication mode** should be specified. In the standard communication, the sending message is copied to a system buffer if the system has enough memory space to copy the data to the system buffer, then the sending process is immediately finished and the process returns to the main program before calling the corresponding receiving function. This means that sending smaller sized messages to the memory space of the system buffer system provides a higher chance for the sending process to finish before calling the corresponding receiving function.

"non-blocking" function in MPI terminology means the receive function returns to program immediately even if the corresponding sending function is not called. The non-blocking function does not guarantee correct sending because some elements of the sending array may be updated during the sending process. To guarantee the correct sending, the user needs to check the status of the sending array with the function **MPI_Wait**. After calling a non-blocking function, computations that are not related to the sending are implemented to hide the time of communications. The readers need to check the implementation of non-blocking function that the communication can start before the computations. This is because some implementations of non-blocking function cannot support starting communication at the time of its calling, and the function only supports communication after calling MPI_Wait. In this situation, no overlapping between communications and computations is established in the user's program.

2.2.2 One-to-One Communication Functions (Procedures)

In this section, one-to-one communication functions (procedures) are explained. The following function is using a standard communication mode.

Receive function (procedure): MPI_Recv

MPI_Recv(recvbuf, icount, idatatype, isource, itag, icomm, istatus, ierr)

The **recvbuf** specifies an address of the receiving array. For example, if we define an array with double precision as "**double A[N]**", we only need to specify the **recvbuf** as "**A**."

The **icount** is an integer type, and it specifies the number of elements for the receiving array. For example, we want to specify the size of 5, a literal "5" can be written, or to specify the number to a variable, say **icount**, by "**integer icount**", then "**icount = 5**."

The **datatype** is an integer type, and it specifies the type for the receiving array. For example, the following types can be specified: **MPI_CHARACTER** (Type of character), **MPI_INTEGER** (Type of integer), **MPI_REAL** (Type of float), **MPI_DOUBLE_PRECISION** (Type of double precision).

The **isource** is an integer type, and it specifies the rank for the process of sending the message. If you want to receive any messages, you can specify the **isource** as "**MPI_ANY_SOURCE**."

The **itag** is an integer type, and it specifies the tag for the message. If you want to receive a message with any tags, you can specify the **itag** as "**MPI_ANY_TAG**."

The **icomm** is an integer type, and it specifies the communicator which can identify a group of MPI processes. Usually, we can specify "**MPI_COMM_WORLD**", which is reserved as a variable for the default communicator, and contains whole MPI processes.

The **istatus** is an array with integer type. This will be returned for information about the received message. The programmer should make sure to allocate the array in advance, and he/she should specify the array size. The array is defined with "**integer istatus(MPI_STATUS_SIZE)**". The rank of sending is stored into "**istatus(MPI_SOURCE)**", and tag number is stored into "istatus(MPI_TAG)."

The **ierr** is an integer type, and it returns the err code of MPI.

Send function (procedure): MPI_Send

MPI_Send(sendbuf, icount, idatatype, idest, itag, icomm, ierr)

The **sendbuf** specifies an address of an array for sending message.

The **icount** is an integer type, and it specifies the number of elements for the sending array.

The **idatatype** is an integer type, and it specifies the type for the sending array.

The **idest** is an integer type, and it specifies the rank for processing the received message.

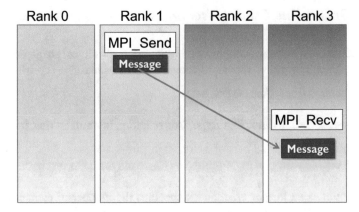

Fig. 2.6 A pattern of one-to-one communication

The **itag** is an integer type, and it specifies the tag for the message.

The **icomm** is an integer type, and it specifies the communicator.

The **ierr** is an integer type, and it returns the err code of MPI.

Let us consider that **MPI_Send** is called by rank #1, and **MPI_Recv is** called by rank #3. The communication pattern in this case is shown in Fig. 2.6.

In Fig. 2.6, the ranks related to this communication are only the rank #1 and the rank #3, while the rank #0 and the rank #2 are not involved in the communications. This type of communication is called "one-to-one communication."

2.2.3 One-to-All Communication Functions (Procedures)

Basically, we can describe any communications with one-to-one communications. However, using only one-to-one communications causes high cost of code managements, such as increasing lines of program. To solve the problem, macro functions for frequently used functions are prepared as APIs in MPI. One of the APIs is a one-to-all communication function for sending one message to all processes. **MPI_bcast** is a typical function of the one-to-all communication function (procedure) and can be defined as follows.

MPI_Bcast(sendbuf, icount, idatatype, iroot, icomm, ierr)

The **sendbuf** specifies an address of an array for **sending and receiving** messages. This is an interesting specification for MPI. As the author mentioned, MPI has only one program that processes sending and receiving, hence it cannot separate sending and receiving arrays. As the result, we need to share the arrays between sending and receiving. The problem is how to identify each process. The answer is by specifying the following arguments.

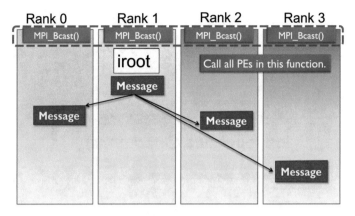

Fig. 2.7 A communication pattern of **MPI_Bcast**

The **icount** is an integer type, and it specifies the number of elements for the sending array.

The **idatatype** is an integer type, and it specifies the type for the sending and receiving array.

The **iroot** is an integer type, and it specifies the rank of communications which has sending messages. The same number should be specified among all processes.

The **icomm** is an integer type, and it specifies the communicator.

The **ierr** is an integer type, and it returns the err code of MPI.

Figure 2.7 shows communication pattern of **MPI_Bcast**.

In Fig. 2.7, the communication does not start until all PEs call **MPI_Bcast**. If one PE does not call **MPI_Bcast**, the other PEs stop in the part of **MPI_Bcast**.

2.2.4   Collective Communication Functions (Procedures)

Reduction operation is a process that reduces its dimension. For example, dot product is a reduction operation, since it reduces the dimensions of a vector (an N dimensional space) to a scalar (one dimensional space). In parallel processing, a reduction operation is one of the types of major operations, needing both communications and computations. Thus, the reduction operation is also called "**Collective Communication**."

In MPI, there are two kinds of functions (procedures) for collective communication with different final results. The first one is:

MPI_Reduce (sendbuf, recvbuf, icount, idatatype, iop, iroot, icomm, ierr)

The **sendbuf** specifies an address of an array for sending message.

The **recvbuf** specifies an address of an array for receiving message. The result is written in the rank specified by **iroot**. The addresses between the sending array and

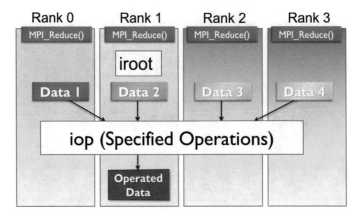

Fig. 2.8 A communication pattern of **MPI_Reduce**

receiving array must be different. The programmer must therefore allocate different arrays between sending and receiving.

The **icount** is an integer type, and it specifies the number of elements for the sending array.

The **idatatype** is an integer type, and it specifies the type for the sending and receiving array.

The **iop** is an integer type, and it specifies the kind of operations. For example, it can specify MPI_SUM (A summation), MPI_PROD (A product), MPI_MAX (maximum), MPI_MIN (minimum), MPI_MAXLOC (maximum and its location), and MPI_MINLOC (minimum and its location), etc.

The **iroot** is an integer type, and it specifies the rank of communications which has sending messages. The same number should be specified among all processes.

The **icomm** is an integer type, and it specifies the communicator.

The **ierr** is an integer type, and it returns the err code of MPI.

With the above arguments, the function (procedure) **MPI_Reduce** operates multiple inputs on each PE by a specified operation, and then a process to obtain a result. Figure 2.8 shows the communication pattern of **MPI_Reduce**.

In Fig. 2.8, the process stops if all PEs do not call **MPI_Reduce**.
The following function (procedure) is another collective operation.

MPI_Allreduce(sendbuf, recvbuf, icount, idatatype, iop, icomm, ierr)

The **sendbuf** specifies an address of an array for sending message.

The **recvbuf** specifies an address of an array for receiving message. The addresses between sending array and receiving array must be different. The programmers must allocate different arrays between sending and receiving.

The **icount** is an integer type, and it specifies the number of elements for the sending array.

The **idatatype** is an integer type, and it specifies the type for the sending and receiving array.

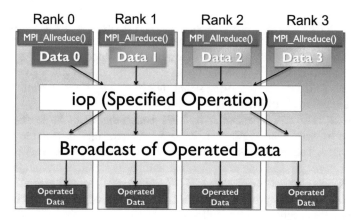

Fig. 2.9 A communication pattern of **MPI_Allreduce**

The **iop** is an integer type, and it specifies the kind of operations.

The **icomm** is an integer type, and it specifies the communicator.

The **ierr** is an integer type, and it returns the error code of MPI.

With the above, the function (procedure) **MPI_Allreduce** operates multiple inputs on each PE by specified operation, and then all processes will obtain the same result. Figure 2.9 shows the communication pattern of **MPI_Allreduce**.

The function of **MPI_Allreduce** can be regarded as a function that **MPI_bcast** is implemented after calling **MPI_Reduce**. Due to the increase of communication to **MPI_Reduce**, **MPI_Allreduce** takes more time than MPI_Reduce. This increase becomes unmanageable the more the number of processes increase. Hence programmers generally avoid many implementations with **MPI_Allreduce**.

2.2.5 An Example of MPI Programming

For an example of MPI programming, a "parallel Hello World" program is shown as the follows:

program main

```
<1> include 'mpif.h'
<2> integer myid, numprocs
<3> integer ierr
<4>    call MPI_Init(ierr)
<5>    call MPI_Comm_Rank (MPI_COMM_WORLD, myid, ierr)
<6>    call MPI_Comm_Size (MPI_COMM_WORLD, numprocs, ierr)
<7>    print *, "Hello parallel world! Myid:", myid
<8>    call MPI_Finalize (ierr)
<9> stop
<10> end
```

There is only one program for MPI, since MPI is based on the SPMD programming model. The readers may feel some incompatibility to sequential programming based on the definition of variables in the line <2>. Understand that the definition (allocation) of a variable is performed with multiple processes simultaneously even in only one definition. This indicates that multiple variables with same name are defined through the MPI execution.

Initialization of MPI is performed by **MPI_Init** in line <4>. The number of ranks is obtained by **MPI_Comm_Rank** in line <5>, then the rank is stored to an integer variable **myid**. The number of MPI processes determined by start-time is obtained by **MPI_Comm_Size** in line <6>, then the number is stored to an integer variable **numprocs**.

Output for the above program depends on the number of processes. If we set it to 4 processes, the output will be as follows.

Hello parallel world! Myid:0
Hello parallel world! Myid:3
Hello parallel world! Myid:1
Hello parallel world! Myid:2

Because we use four processes, the four output lines are given by the print statement. If we set 1000 processes, we will obtain 1000 lines of output. The numbers printed are different, since the output is for the number of ranks, and each number of ranks must be a different number in MPI. In addition, the number is not sorted—this tells us that the execution is not synchronized. This also indicates that the order of the printed numbers is different than in the actual execution. Please note that the order in parallel execution will be different from order shown for the ranks, since the output comes from orders to write for a standard output buffer.

2.2.6 Consideration of Communication Algorithms

Now, let us consider a collective communication in which the result is stored to a process that adds distributed data over all processes. This function can be described with only one line with **MPI_Reduce**, but we will describe the implementation of **MPI_Reduce** with a one-to-one communication.

The first implementation is known as "**Sequential Sending Method**." In the sequential sending method, it starts from rank #0, then the rank #0 sends data to rank #1. In rank #1, its own data is added with the data received from rank #0. After the addition, rank #1 sends the data to rank #2, and the process is repeated. Finally, the result from rank #$p - 1$ is obtained when we use p processes.

Obviously, the sequential sending method has no parallelism for the communication and summation parts. By this reason, we use another method, which is known as "**Binary-Tree Sending Method**", or "**Tournament Method**" to utilize parallelism for the parts of communication and summation. The concept of binary-tree sending method is as follows. If we operate a sequence of "$1 + 2 + 3 + 4 + 5 + 6 + 7 + 8$", we apply the following process if we can use 4 processes and each process has two items:

- Phase 1

  Rank #0: $1 + 2 = 3$, then send it to rank 1,
  Rank #1: $3 + 4 = 7$,
  Rank #2: $5 + 6 = 11$, then send it to rank 3, and
  Rank #3: $7 + 8 = 15$.

- Phase 2

  Rank #1: $3 + 7 = 10$, then send it to rank 3, and
  Rank #3: 3: $11 + 15 = 26$.

- Phase 3

  Rank #3: $10 + 26 = 36$.

As the author had shown the above, we can parallelize the process if we can apply a change of computation orders for the summation. This example also shows that we need to consider a change of parallel algorithms to obtain much parallelisms.

Comparing the complexity for communications and computations for both algorithms, sequential sending method requires (*nprocs* − 1) times, while binary-tree sending method requires $\log_2$ (*nprocs*) times, for communications, respectively. Hence binary-tree method is superior to sequential sending method, but this complexity does not include the effect of network topology.

The assumptions for the above cost estimation for communications are based on the parallel sending in each phase. This indicates that there is no **message collision**, and/or **traffic congestion** for the sending. In actual communications, message collisions will occur when several processes perform communication simultaneously. The collision depends on allocations of MPI processes to nodes and the topology of the network. In addition, the size of the messages affects the communication; the above example is only sending a message with double precision. Imagine a case that a 1 GB message is sent to all processes simultaneously. This is a totally different situation to the example. Moreover, we need consider the size of parallelisms, such as 2 parallels vs. 20,000 parallels. As a conclusion of this discussion, binary-three sending method is not always faster than sequential sending method. It is dangerous to evaluate parallel performance only by using theoretical analysis. We need to evaluate performance with actual problems and actual parallel computers.

Communications like **MPI_Bcast** are performed inside of **MPI_Allreduce**. Performance of **MPI_Bcast** affects the total performance. The major implementation of MPI_Bcast is based on a binary-tree sending method in general. Hence the readers are advised to understand the inner workings before calculating the parallel performance.

2.2.7 Allocation of MPI Processes

In the final part of this chapter, the allocation between MPI processes and nodes are explained. Hereafter, we call this allocation "**MPI node allocation**." In MPI, the MPI node allocation is performed by a user-defined file, usually named "**Machine File**".

In supercomputing environments, the MPI node allocation is performed by a batch job system automatically. Hence there may be a limitation of MPI node allocation. In the case that batch job system performs MPI node allocation, it may be difficult to optimize the allocation with respect to network topology. In the worst case, message collisions are caused frequently by the MPI node allocation, which reduces the communication performance. Some batch job systems provide a way to control MPI node allocation to users by letting them specify the network topology themselves.

Another issue for MPI node allocation is the policy of operation by supercomputer centers. To maximize usage of nodes, some supercomputer centers do not provide a function that users can specify network topology with, or do not provide exclusion of other jobs on nodes. This implies that MPI process is allocated to non-continuous nodes, or shared nodes with other jobs to maximize node usage and to increase job throughputs. These cases increase the communication time due to message collision or traffic congestion. We need to consider some modifications in communication implementations to reduce communication time imposed by the limitation of operation for computers.

Exercises

1. What is the percentage that needs to be parallelizable in the case of execution with 10,000 cores and 99% of more effectiveness of speedups? Use Amdahl's Law to compute the value.
2. Survey MPI communication modes and their MPI functions (procedures).
3. Implement sequential sending method and binary-tree method with a parallel computer installed in MPI. Evaluate the performance between them by varying the amount of messages from 1, then find the best method with respect to the amount of messages. Does the best method change according to the amount of messages?

References

1. Message Passing Interface Forum. http://www.mpi-forum.org/
2. P.S. Pacheco, *Parallel Programming with MPI* (Morgan Kaufmann, Burlington, 1996)

Chapter 3
Basics of OpenMP Programming

Takahiro Katagiri

Abstract This chapter explains the basics of OpenMP programming, which is widely used in parallel programming language for thread parallelization with inside node parallelism. Specifications of major constructs and clauses for OpenMP are explained with several examples of programming. Scheduling, which is a job allocation method for OpenMP, is also explained. Code optimization techniques for OpenMP, named First Touch, is also shown. Finally, the possibility of extension to Graphics Processing Unit (GPU) from programming with OpenMP is explained with OpenMP version 4.0 and OpenACC.

3.1 What Is OpenMP?

3.1.1 Overviews

OpenMP (OpenMP C and C++ Application Program Interface Version 1.0) [1, 2] is a specification that contains (1) Directives; (2) Libraries; and (3) Environmental variables; for parallel programming with shared memory machines. Programmers describe parallelization with directives provided by OpenMP specification.

The most important thing when using OpenMP is that OpenMP does not provide automatic parallelization. Programmers need to parallelize their codes with OpenMP directives by themselves.

It is said that programming with OpenMP is easier than programming with MPI since programming with OpenMP has no data distribution and modifying for loop variables such as modifying starting and ending indexes and no explicit description of communications. However, it is also known that using OpenMP is difficult and increases the costs of implementation to establish high performance. In this point, there is believed to be no difference for developing cost to establish high performance

T. Katagiri (✉)
Information Technology Center, Nagoya University, Nagoya, Japan
e-mail: katagiri@cc.nagoya-u.ac.jp

© Springer Nature Singapore Pte Ltd. 2019
M. Geshi (ed.), *The Art of High Performance Computing for Computational Science, Vol. 1*, https://doi.org/10.1007/978-981-13-6194-4_3

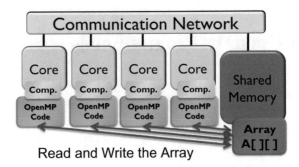

Fig. 3.1 Overview of thread executions by OpenMP. Multiple accesses with multiple cores are made. Hence it requires the use of mutual exclusion for accesses to shared memory to keep the same results in sequential processing

between OpenMP and MPI. Currently, it is almost impossible to obtain more than 100 speedups by using OpenMP.

In Fig. 3.1, an overview of processes for OpenMP is shown.

In Fig. 3.1, parallelization with OpenMP makes threads. Each thread is allocated to each core. Each thread accesses the shared memory simultaneously. Hence, uncontrollable updates may happen, and so we need to use mutual exclusion for the accesses to keep sequential results.

3.1.2 Target Parallel Computers

The target parallel computers for OpenMP are the shared memory type. Hence if OpenMP is applied to a distributed memory type, it can only parallelize the inside node. As of 2017, parallelism in the inside node is around 200 times. Therefore, the only way to obtain massive parallelisms is to use MPI.

On the other hand, by adapting parallelization with OpenMP for the inside node, and by adapting parallelization with MPI, we can reduce the total number of parallelism with a factor of 1/200 by using full cores in the system. This reduces communication time since a reduction in the total number of parallelisms can contribute to the reduction of communication time. This style of execution is called hybrid MPI/OpenMP execution. The hybrid MPI/OpenMP execution is one of the promising ways to establish high performance on the future massively parallelized systems.

3.1.3 How to Compile

We can compile codes with OpenMP by specifying a compile option for OpenMP parallelization, which is usually provided by compilers supported by OpenMP.

As the author will mention later, parallelization with OpenMP is done by specifying a comment, which is called "directive." The execution can be performed sequentially if programmers do not like parallel processing by specifying compiler option for sequential processing. This is a great merit for OpenMP because just using one program is enough between thread parallel processing and sequential processing.

The readers should note that loops without OpenMP descriptions are sequential; OpenMP is not an automatic parallelization. Some compilers, however, do support automatic parallelization. In this case, thread parallelization by OpenMP and by the automatic compiler are also adaptable simultaneously. This supports parallelization that loops with OpenMP are parallelized according to directives by OpenMP, and the other loops are automatically parallelized by the compiler.

3.1.4   Descriptions for Parallelization with OpenMP

As the author mentioned, parallelization by OpenMP is specified by directives. The directives are defined by languages, such as C or Fortran. In Fortran,

Statements that start with "**!$omp**"

are the way to specify parallelization by OpenMP.

An execution method after compiling OpenMP program is to execute the "*executable code*" directly. During this time, users can specify the number of threads by specifying an environmental variable, named "OMP_NUM_THREADS." The following is an example for an executable code "**a.out**." In addition, this is an example that shell of Linux is bash.

$ export OMP_NUM_THREADS = 16
$ ./a.out

Here, execution time between sequential execution and parallel execution with **OMP_NUM_THREADS = 1** may be different. This is because some overheads, such as control statements of OpenMP, may be inserted by the computer for the OpenMP program. Hence, sequential execution is faster than sequential execution with compared to thread parallelized code by OpenMP in usual. Considering this fact, it is unfair to evaluate the speedup factor by using sequential execution with thread parallelized code by OpenMP. Unfortunately, it is thought that many evaluations are performed under this kind of situation.

Please note that the following example for Fortran programs is based on free format due to the restriction of pages, although free format and fixed format are available for OpenMP.

3.2   Programming Modes of OpenMP

3.2.1   Parallel Construct

One of the frequently used directives in OpenMP is parallel construct. Figure 3.2 shows the parallel construct.

Figure 3.2 shows a program that executes the following sequence: "**Block A →**
Block B → Block C", where the **Block** indicates a unit of the program, such as a loop structure or a cluster of sentences. We apply parallel construct to **Block B** in Fig. 3.2. Then **Block A** is executed sequentially, but after finishing the execution, **Block B** is executed in parallel on each thread. After finishing **Block B**, the threads are joined, and **Block C** is then executed.

3.2.2   Work Sharing Construct

For a parallel construct like the one shown in Fig. 3.2, part of the parallel processing that is described by directives in OpenMP (such as **Block B** in Fig. 3.2) is called "**parallel region.**" A construct of OpenMP that can specify a parallel regions and execute the process in parallel is called "**work-sharing construct.**"

There are two kinds of work sharing constructs:

(1) To specify it inside the parallel region: These are: **do construct, sections construct**, **single construct** (or **master construct**), etc.
(2) To specify it in combination with parallel construct. There are: **parallel do construct, parallel sections construct**, etc.

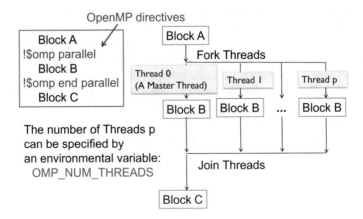

Fig. 3.2 An example of parallel construct

3.2.3 Parallel Do Construct

One of the well-used directives of OpenMP is **parallel do construct**. Figure 3.3 shows the explanation.

In Fig. 3.3, the process is divided by each thread of OpenMP to allocate the same loop length for the loop **i**, then the process (inside the statement of the loop) is executed in parallel.

The most important thing to note when using parallel do construct is that the application of parallel do construct is limited; it cannot be applied to arbitrary loops. The correct case to apply parallel do construct is shown in Fig. 3.3, when we want to return the same result from sequential execution. The compiler does not know whether the loop can or cannot be adapted for parallel do construct. Additionally, there is no warning message if the loop is not adaptable in some compilers. Hence again, the programmers have to decide when to apply the parallelization using the parallel do construct.

With the above discussions, the readers can understand that OpenMP is not an automatic parallelization compiler. Later, the author will explain the kinds of loops that can be parallelized by OpenMP.

In the example in Fig. 3.3, the lengths of the loops are equally allocated to each thread with respect to the number of threads defined by the user at run time. This is the default execution in parallel, but programmers can change the approach. The loop length to allocate each thread or the method to allocate the loops is called "**scheduling**." Scheduling in OpenMP will be covered later.

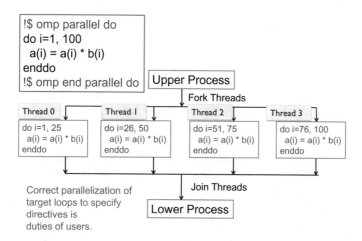

Fig. 3.3 An example of parallel do construct

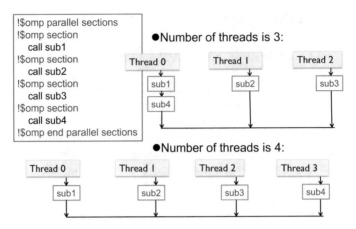

Fig. 3.4 An example of sections construct

3.2.4 Sections Construct

In the case of each thread having different jobs, **sections construct** should be implemented. Figure 3.4 shows an example of sections construct.

In Fig. 3.4, each process between the directives "**!$omp section**" is allocated to each thread, and is executed in parallel. The procedures from **sub 1** to **sub 4** can be executed in parallel according to the number of threads in the example in Fig. 3.4. No dependency between the procedures **sub 1** to **sub 4** is needed.

3.2.5 Critical Construct

When we parallelize programs with OpenMP, some parts need to perform sequential executions. The directive that creates a region specified by directive of OpenMP that will be performed sequentially (which indicates that only one thread can enter the region) is called **critical construct**.

Critical construct provides a region that is specified between "**!$omp critical**" and "**!$omp end critical**" and only one thread can enter the region in parallel execution. This region is called "**critical section.**" The result in parallel execution is the same as that of sequential execution.

The readers should be careful about the reduction of performance when doing critical construct. Since implementing critical construct frequently checks whether each process enters a critical section or not, the cost of checking is very high. In particular, the execution time of the checking increases according to the increase of the number of threads. Hence, total performance goes down if we use too many critical constructs. It is better not to use the critical construct if possible.

3.2.6   Private Clause

Variables in directives of OpenMP, such as in parallel do construct, are shared variables by default. The shared variables in OpenMP are shared in each thread. If the shared variables are accessed simultaneously in each thread, the result will not match with the result for the sequential execution.

Let us consider an addition "$i = i+1$", performed in each thread with a shared variable "i." In this addition, the result between sequential and parallel execution may be different. This is because of the possibility that when a thread has just been stored to the shared variable, other thread has the read data from the shared variable. To match the result between sequential and parallel executions, we need to use critical construct.

Note that loop induction variables are also shared variables in OpenMP. Let us consider a loop "**do i = 1, n**". The loop induction variable "i" is also a shared variable. Hence, it cannot count the number of loops elapsed even if we use parallel do construct[1].

How can we resolve this problem? The answer is to use a directive to specify a variable privatization for the specified variables in each thread. **Private clause** can perform the variable privatization in OpenMP.

For example, when we apply parallel do construct to the outer loop in the following two-nested loop:

```
do i=1, n
  k = a(i)
  do j=1, n
    b(j) = b(j) + k * c(j)
  enddo
enddo
```

the following code should be described:

```
!$omp parallel do private(k,j)
do i=1, n
  ...
!$omp end parallel do
```

---

[1] An exception is the loop that occurs just after the parallel do construct. The loop induction variable in the loop becomes a private variable automatically. Some compilers provide automatic variable privatization of loop induction variables, but this is against the specification of OpenMP. Hence, the readers need to realize that the variable privatization is a compiler specific process.

3.2.7   Reduction Clause

When we parallelize a code, it may occur that the result needs to be added between temporal results in each thread. We call the processes that need parallel executions and collaborative operations between each thread "**reduction operation**."

To implement the reduction operation in OpenMP, **reduction clause** is prepared.

Reduction clause provides addition with "**+**", subtraction with "**−**", multiplication with "*****", maximum with "**MAX**", and logical AND with "**.AND.**", and so on.

For example, in the following code, we adapt parallel do construct and we adapt reduction clause to a shared variable named "**s**":

```
do i=1, n
   s = s + a(i)
enddo
```

we can describe the code as follows:

```
!$omp parallel do
!$omp& reduction(+ : s)
do i=1, n
   s = s + a(i)
enddo
!$omp end parallel do
```

3.2.8   Other Constructs

In this section, we explain other constructs.

Single Construct

For specific parts that should be sequentialized in parallel do construct, "**single construct**" can be used.

The region from "**!$omp single**" to "**!$omp end single**" is executed in a thread with exclusive control. Following that, synchronization among threads is performed. Hence, additional synchronization time is required in single construct.

If synchronization is not required, we can remove it by specifying "**!$omp nowait**" just after the single construct.

Difference between single construct and critical construct is: the single construct is performed with the part for only the first thread that enters, while the critical construct is performed with the part for all threads.

Master Construct

There is a case that **master construct** is faster than single construct. The reason is because the master construct performs the part that is specified with the directive of master construct (which is a region with **!$omp single—!$omp end single** in the previous example) should be executed in the master thread, which is usually thread numbered #0. In addition, there is no synchronization after finishing the region for the master construct.

Flush Construct

Flush construct makes explicit consistency values between variables and physical memory. Variables specified by flush construct are made consistent in specified locations. The other variables are not the same as the values stored on physical memory in parallel execution. This implies that computed results are only stored in registers. Hence, to obtain a computed result, specifying flush construct to store back to physical memory for the result is required.

The following constructs are inserted with flush after finishing, i.e., we need no explicit flush construct: the input and output parts of **barrier construct** and critical construct, the output part of the parallel construct, sections construct, and single construct.

The readers should also note the reduction of performance when using the flush construct, since the performance goes down if you use too many flush constructs. Description of flush construct is as follows:

!$omp flush (List of target variables)

where all variables are target if the list of the target variable is omitted.

3.2.9 Scheduling

As explained this in the section of parallel do construct, default scheduling in OpenMP is dividing loop length by the number of threads to allocate computations with the equal division to each thread. This type of scheduling works well when computations to be allocated in loop length can be divided equally. However, general loops cannot be assured of such "equal division" for computations. If each computation for loop iteration causes any imbalance, such kind of scheduling (shown below) will not work. As a conclusion, the efficiency of thread parallelization decreases. This situation is called "**load imbalance**."

The above situation is explained in Fig. 3.5.

> In **parallel do** construct, it divides space of loop iterations
> equally by number of threads to make continuous loop.
> Then it makes parallel processing.

(a) A Default Division.

> In this situation, if computation loads is not balanced,
> we cannot obtain enough speedups by parallel execution.

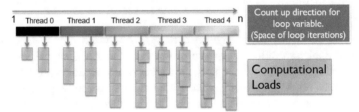

(b) A Case of Load Imbalance.

> To balance the computational loads, length of allocation is
> shortened, and the loads allocate cyclic manner.

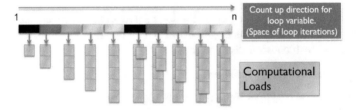

> The best length of allocation, we call this "chunk size",
> depends on computer architectures and target processes.
> To do the above allocation, OpenMP provides a construct.

(c) A Case of Improvement of Load Imbalance.

Fig. 3.5 An explanation of loop scheduling

In Fig. 3.5, the computations for each loop iteration increase in complexity due
to the increase of loop counts. In this situation, load balancing cannot be obtained
by using the default division of loops in Fig. 3.5 a. However, if the jobs are allocated
with a length of blocks, load imbalance issue can be tackled. The length of blocks is
called "**Chunk Size**."

> **schedule (static, _n_)**
>
> > This scheduling divides loop length by chunk size, and it allocates the load from Thread 0 as Round-robin method with cyclic manner. The chunk size can be specified in the above _n_.
>
> > Without specifying schedule construct, static is default scheduling, and chunk size is set to loop length / number of threads.

Fig. 3.6 An explanation of the **static** clause

The chunk size to allocate jobs into each thread is calculated in advance for the default division of OpenMP. This method is called "**Static Scheduling**." The clause to specify static scheduling for OpenMP is "**static clause**." Figure 3.6 shows an explanation of the static clause.

In Fig. 3.6, "**static**" is described in static clause "**schedule()**." The chunk size is described in "**n**" in the "**schedule()**." Since the optimal chunk size depends on processes in loops and hardware, we will not be able to know what the size is without measuring the performance in advance. In this way, the chunk size is also a tuning parameter.

On the other hand, we can use "**dynamic scheduling**" if the jobs cannot be allocated in advance. In dynamic scheduling, the jobs are allocated to threads with the length of the same chunk size as static scheduling in the first phase. However, after finishing the first phase, the next job is allocated to the fastest thread that has finished the allocated job in the first phase in a first-come, first-served basis. The dynamic scheduling is described with "dynamic" in dynamic clause "**schedule()**." An explanation of dynamic clause is shown in Fig. 3.7.

In Fig. 3.7, a chunk size needs to be specified in the **schedule()**. The chunk size is also a tuning parameter.

> **schedule(dynamic, _n_)**
>
> > This scheduling divides loop length by chunk size, and allocates process to thread that finishes process in first-come, first-served basis. The chunk size can be specified in the above _n_.

Fig. 3.7 An explanation of **dynamic** clause

Even in dynamic clause, the chunk size is a constant number. If we need to change the chunk size dynamically in the loop, we will need a different clause. The clause that can vary the chunk size according to loop iterations is called **guided clause**.

The above scheduling is described at the end part of the constructs, such as parallel do construct.

3.3 Some Notes on the Programming

A frequently used construct in OpenMP is parallel do construct. One of the things we must do for parallel do construct is to establish private clause. The private clause is needed to specify all variables to allocate them in each thread privately. As the author mentioned, default definition in OpenMP is shared variables except for those explicitly defined with private clause set by users. In addition, compilers print no warnings; hence, many programmers cannot find this problem. Things go out of control when specifying too many private variables, for example, the number of target variables is more than 100 in actual programs in each loop. Hence, the cost of specification for private clause is not small.

3.4 First Touch to Speed up for OpenMP Programs

As the author mentioned in Chap. 1, many current computer architectures are based on NUMA. In NUMA, speeds of access to memory are different between neighbor memory and far memory, even though both are shared memories. In the NUMA architecture, programmers can put allocated arrays into the near memory by software control.

"**First Touch**" is known for an optimization method for caches, in particular, for **ccNUMA** (Cache Coherent Non-uniform Memory Access) architectures. First Touch is one of the code optimization techniques for OpenMP programming on ccNUMA architectures. In the architecture of ccNUMA, allocated arrays are set to the nearest core when the first access for the arrays is performed. By using this mechanism, we implement initialization processes for the allocated array to do the same computations for the target kernel (this is also same loop structure in many cases) in the first part of the program by using parallelization with OpenMP. After the initialization, the allocated arrays are set to the nearest memory.

The following is an example of First Touch.

!The following is an initialization loop for the array A and B. This allows us to set the array A and B to the nearest memory.

```
!$omp parallel do private( j )
do i=1, 100
  do j=1, 100
    A( i ) = 0.0d0
    B( i , j ) =0.0d0
  enddo
enddo
!$omp end parallel do
...

! The follow is main loop. The arrays A and B,
which are on the nearest memory, are accessed in the loop.
!$omp parallel do private( j )
do i=1, 100
  do j=1, 100
    A( i ) = A( i ) + B( i , j ) * C( j )
  enddo
enddo
!$omp end parallel do
```

3.5 Extends to GPGPU

Recently, specification of **OpenMP version 4** (Hereafter, we denote OpenMP 4.0) [3] is defined. In OpenMP 4.0, directives to utilize Graphics Processing Unit (GPU) are specified. The programming with GPU is also called "**General Propose GPU (GpGPU)**". OpenMP 4.0 also can be used for a programming language to do GpGPU.

It is easy to translate a program from OpenMP parallelized code for multicore CPUs or many-core CPUs to GPU parallelized code by only specifying some directives. For example, GPU is specified as a target device in OpenMP 4.0, the following **target construct** can be used:

```
!$omp target
do i=1, n
    A(i) = A(i) + B(i)
enddo
!$omp end target
```

For the above arrays A and B, data is automatically transferred from memory to device memory on GPU. Since data movement happens every time when calling target construct for all arrays inside the loop of the target construct, too many transfers from memory to device memory on GPU, or vice versa, are performed. As a result, the performance goes down in this situation. To obtain high-performance, we should reuse the data on device memory on GPU to minimize data movement between memory and device memory on GPU.

To reduce the data movement between memory and device memory on GPU, there is a method to specify dedicated arrays to send data to device memory on GPU. **Map clause** is one way to specify this. For example, when array **A(10)** on memory is specified with map clause to transfer to device memory on GPU, the description of this process is:

```
!$omp target data map(A)
    Computation with A().
!$omp end target data
```

As mentioned above, the main topics of programming with OpenMP 4.0 are considering minimization of data movement between memory and device memory on GPU.

Unfortunately, OpenMP 4.0 is not well used for programming, and its performance is not satisfactory and not well evaluated with respect to the cases of programming with **CUDA**, which is one of major computer languages for GPU.

On the other hand, **OpenACC** [4, 5], which is another directive-based programming interface for GPU, can also be used. Compilers for OpenACC also provide automatic translation to CUDA. The principal difference between OpenMP 4.0 and OpenACC is thought to be negligible. Minimization of data movement from memory to device memory on GPU is the most important and common issue for performance even in OpenACC. The description of the data movement in OpenACC is **data clause**.

As mentioned in the last section, there are many computer languages to establish easy programing for GPU now. Programming for GPU is predicted to become friendlier for many programmers in the very near future.

Exercises

1. Verify that the result does not match between sequential execution and parallel execution if we do not use reduction clause when making a program.
2. Parallelize a code for matrix-matrix multiplications with OpenMP. Evaluate speedups based on one thread execution and sequential execution. Do you find any difference between them?
3. Evaluate speedups for the parallelized code for matrix-matrix multiplications with First Touch. Do you find speedups by applying the First Touch?
4. Survey specifications for OpenMP version 4.0.
5. Survey specifications for OpenACC version 1.0, version 2.0.

References

1. OpenMP Forum. http://openmp.org/
2. OpenMP Fortran Application Program Interface (1997), http://www.openmp.org/mp-documents/fspec10.pdf. Oct 1997
3. OpenMP Application Program Interface Version 4.0 (2013), http://www.openmp.org/mp-documents/OpenMP4.0.0.pdf. July 2013
4. OpenACC Home. http://www.openacc.org/
5. OpenACC 2.0 Specification. https://www.openacc.org/sites/default/files/inline-files/OpenACC_2_0_specification.pdf

Chapter 4
Hybrid Parallelization Techniques

Takahiro Katagiri

Abstract This chapter explains the techniques of hybrid parallel execution to establish 10,000 more parallel executions and reduce communication time. Several terms are defined in the discussion of hybrid parallel execution. In addition, actual examples of hardware and programming for parallel execution are shown. Finally, an experimental methodology to develop hybrid MPI execution is shown.

4.1 Overview

In the first part of this chapter, we summarize the terminology of hybrid parallel execution as follows:

- Pure MPI execution: Only MPI is used for parallel programs.
- "MPI + X" execution: MPI and some component X are used for parallel programs, where X is defined with a wide meaning in general and X can be OpenMP, an automatic parallelization with compiler, dedicated languages for GPU, such as CUDA, or other many-core languages, such as OpenACC, including their combinations. These uses as X depend on the current mainstream computer architectures.
- Hybrid MPI execution: Parallel programs use OpenMP (called Hybrid MPI/OpenMP execution) or GPU as X of MPI+ X execution.

Let us consider using the same resource amount (the same total number of cores). In this situation, the number of processes for MPI is bigger than that of Hybrid MPI execution. This also implies that using Hybrid MPI execution reduces communication time because of the reduced number of MPI processes. Hence, the main aim of introducing hybrid MPI execution is to reduce communication time in massively parallel execution, such as more than 10,000 parallelisms in practical.

T. Katagiri (✉)
Information Technology Center, Nagoya University, Nagoya, Japan
e-mail: katagiri@cc.nagoya-u.ac.jp

© Springer Nature Singapore Pte Ltd. 2019
M. Geshi (ed.), *The Art of High Performance Computing for Computational Science, Vol. 1*, https://doi.org/10.1007/978-981-13-6194-4_4

4.2   Target Computer Architectures

Hybrid MPI execution is a parallel programming model to fit current trends of computer architectures. To understand the merit of parallel programming, we need to know the trends in computer architectures. Figure 4.1 shows an example of the current computer architectures. This is the T2K Open Supercomputer (U. Tokyo) (the HITACHI HA8000 Cluster System), which was installed in Information Technology Center, The University of Tokyo. Its node uses the AMD Quad-Core Opteron (2.3 GHz).

In Fig. 4.1, four CPUs are organized in this architecture. Each CPU is called "**Socket**". Each memory is allocated to near socket. The far memories can be accessed by arbitrary sockets; hence, the node is organized as a shared memory. For access time for memories, the access time for the nearest memory is faster than that of the far memories. Figure 4.2 shows this explanation.

"NUMA" organization is mentioned in Fig. 4.2 as well as "Cache Coherent NUMA (ccNUMA)" organization that provides consistent data on cache memory in the shared memory. The T2K Open Supercomputer (U. Tokyo) in Fig. 4.2 is ccNUMA organization.

Figure 4.2 indicates that we can obtain speedups to allocate arrays on the nearest memory in a parallel program. Hence, it is optimal to allocate an MPI process to each socket from #0 to #3 with four MPI processes, as shown in Fig. 4.2. Each MPI process involves four threads inside the allocated socket and uses the nearest memory. This is the best parallel execution in terms of computer architectures shown in Fig. 4.2.

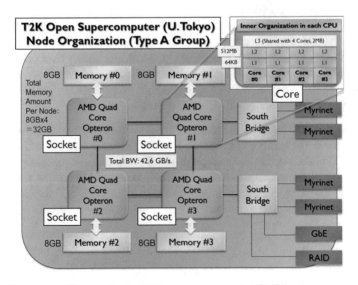

Fig. 4.1 Computer architecture for the T2K open supercomputer (U. Tokyo)

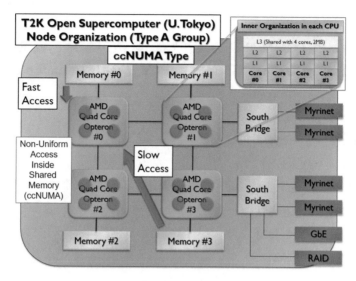

Fig. 4.2 Differences of memory access on ccNUMA organization

Since the hybrid MPI execution is currently mainstream for parallel programming, programmers need to consider the best execution style for it.

As of 2017, each node has around 2–4 sockets, and each socket has fewer than 100 cores. Due to the nature of ccNUMA organization, the performance of memory access goes down, but it can provide a large amount of memory. For example, the SGI UV2000 in the Information Technology Center, Nagoya University [1] provides a total of 20 TB shared memory with a 128 GB memory socket connected to 160 sockets.[1]

4.3 Examples of Execution

In this section, actual examples of hybrid MPI execution are shown.

A hybrid MPI execution program to add OpenMP thread execution to MPI program for matrix-vector multiplications is shown.

---

[1]Performance goes down dramatically if a programmer does not carefully think about optimization of memory access due to frequently far memory accesses on the ccNUMA organization. Hence, even if users can use such a machine with a large amount of memory, the users may not obtain the desired performance of parallel execution.

```
<1> call MPI_INIT(ierr)
<2> call MPI_COMM_RANK(MPI_COMM_WORLD, myid, ierr)
<3> call MPI_COMM_SIZE(MPI_COMM_WORLD, numprocs, ierr)...
<4> ...
<5> ib = n/numprocs
<6> jstart = 1 + myid * ib
<7> jend = (myid+ 1 ) * ib
<8> if ( myid .eq. numprocs- 1 ) jend = n
<9> !$omp parallel do private(i)
<10> do j = jstart, jend
<11>   y( j ) = 0.0d0
<12>   do i= 1 , n
<13>     y( j ) = y( j ) + A( j, i ) * x( i )
<14>   enddo
<15> enddo
<16> !$omp end parallel do
```

In the above program, the loop length which is given by dividing outermost j-loop from 1 to **n** by the number of MPI processes, **numprocs**, is allocated to each MPI process in lines <5>–<8>. Each MPI process executes parallel processing with OpenMP for loop length from **jstart** to **jend** inline <10>.

With the above program, we can observe an interesting aspect. The loop length of parallel processing from **jstart** to **jend** is shortened as the MPI processes increase. If the loop length from **jstart** to **jend** becomes short, the number of threads that OpenMP can execute is limited. In theory, if the loop length is set to 10, there should be no parallelism of more than 10. In current many-core processors, there are more than 200 cores inside a CPU. Hence, the loop length should be more than 200, hopefully, several times more with respect to the prefetching data to optimize memory accesses and for the loop length on each MPI process to obtain high performance. We need to consider reconstruction of algorithms to obtain higher parallelism for current many-core type CPUs.

4.4 Basics of Parallel Programming for Pure MPI Execution

As mentioned in the previous section, we need to parallelize programs with MPI to adopt hybrid MPI execution. To perform this, pure MPI programming should be first adapted to each program.

Due to the page restriction of this chapter, please refer to other references on how to develop pure MPI programming in details, such as Ref. [2]. In this section, an overview to develop pure MPI programming is discussed.

To develop a pure MPI program, developing a sequential programming is very important in most cases. In terms of debugging, it is not good for starting with MPI programming during the first phase. Hence, the most important thing is to make a sequential program without bugs, and make a correct program to obtain the correct output. With this approach, the following methodology is used to make pure MPI program.

1. Make a correct sequential program.

- Make a simple MPI program.

2-1. Divide each process with respect to the computational functions inside the algorithm used, and then gather the correct output before and after the functions.
2-2. Make a MPI program based on item 2-1 in each divided function. In this phase, we do not perform data distribution for MPI.
2-3. Debug the program with verified results between items 2-1 and 2-2 for the sequential program and the parallel program to each function.

- Make a full MPI program.

3-1. Using the final version of the simple MPI program in item 2-3, data distribution can now be implemented for the program.
3-2. Debug a full MPI program in each function with the sequential outputs in item 2-1.

In the above methodology for parallelization, items 2-1–2-3 of making a simple MPI program are the key points. It is impossible to make a full MPI program with full data distribution, even with a highly experienced programmer for MPI programming. The main reason for this is the difficulty to separate the programming issues between algorithm and parallelization. As a result, the total cost of developments increases if we apply the methodology shown above.

Problem size cannot increase for the simple MPI program in item 2 since memory allocation is done in a sequential manner, thus the required memory area is not distributed. However, it can omit implementations of communication and/or change of indexes for arrays for parallel processing. Hence, programmers only change loops of MPI process for the target of parallelization. This modification is shown in the example shown above for the processes in lines <5>–<8> and <10>.

By using the developed simple MPI program, the total cost of developments can be reduced because programmers can concentrate on program parallelization for data distribution.

In some cases, it is possible to finish parallelization only by making the simple MPI program. In this case, the application requires parallelization of a huge amount of computations, but not for large memory space. It is thought that the cost of development for simple MPI programming is the same as that of OpenMP programming. Therefore, the cost of MPI programming is not always much higher than that of OpenMP.

4.5 Basics of Parallel Programing for Hybrid MPI Execution

To execute hybrid MPI programming, we should start by making a pure MPI program. The following is a methodology for hybrid MPI programming.

1. Make a correct pure MPI program.
2. Parallelize the pure MPI program with OpenMP.
3. Evaluate the performance for the program of item 2.
4. If the performance evaluation in item 3 is not satisfied, tuning for OpenMP should be performed.
5. Return to item 3.
6. Evaluate the total performance. If there is lack of performance in communication parts, tuning for communications is performed.

If users want to parallelize program without OpenMP and want to utilize GPUs, then they need hybrid MPI programming with GPU, or CUDA, OpenMP 4.0 or OpenACC, which are directive-based languages for GPU processing.

Recently, MPI and OpenMP are frequently used for CPUs, and they are also used for GPUs, which indicate that both CPU and GPU are used simultaneously as hybrid MPI execution. In this case, we can adapt to the above methodology. However, the hybrid MPI execution may be increased for communication time from GPU to GPU by MPI communications, if there is no hardware support for GPU, like NVLINK [3]. It may be necessary to implement efficient communications or to change the algorithms used.

4.6 Some Notes to Hybrid MPI Programming

In this section, the author makes some notes in hybrid MPI programming for beginners.

4.6.1 Wrong Approach for Parallelization

By spreading multi-core type CPUs, there is a big chance of making MPI programs based on already developed OpenMP programs. In this situation, we need to parallelize loops of **parallel** construct by OpenMP to make an MPI program. If users make parallel loops of MPI, which are inside the loops of OpenMP, the execution time will extremely increase due to many communications, or in the worst case, it cannot run due to the overflow of communication buffer.

4.6.2 Method to Allocate Cores for MPI Processes

No effect on hybrid MPI execution is obtained if the allocation between physical socket and MPI processes are not in the desired manner. Let us consider a parallel computer with 16 cores for each socket, and 4 sockets on a node. In a parallel execution with four MPI processes for a node in this computer, we need to allocate one MPI process to each socket to access the nearest memory for OpenMP threads, which are invocated by each MPI process to obtain high performance. We can think of this as the example of T2K Open Supercomputer (U. Tokyo) in Fig. 4.1. If we allocate four MPI processes to socket #0 in Fig. 4.1, then we obtain no effect of hybrid MPI execution since many far memory accesses happened.

For the allocation of the physical socket, we can use "**numactl**" command at execute-time for Linux. Some supercomputer environments provide a dedicated way to perform this allocation.

4.6.3 Numerical Libraries and Hybrid Parallelization

Some numerical libraries support hybrid MPI parallelization. If numerical libraries are parallelized with threads, we can make hybrid MPI program with thread parallelized numerical libraries.

ScaLAPACK [4], which is a standard numerical library for dense matrices, usually supports hybrid MPI execution. This is because, ScaLAPACK is parallelized with MPI, and ScaLAPACK is implemented with LAPACK [5], which is a standard sequential numerical library. LAPACK is implemented with BLAS [6], which are basic linear algebra subprograms. BLAS support thread execution. Because the lower layer of BLAS is supported with threads, LAPACK, which is a top layer of BLAS, can perform thread executions. Hence, ScaLAPACK, which is the top layer of LAPACK, can perform hybrid MPI execution.

4.6.4 Relationship Between Parallelization and Compiler Optimizations

To adapt MPI parallelization or OpenMP parallelization, construction of loops should be modified from a sequential manner to a parallel manner. When the modification is adapted, some compilers generate inefficient codes. This implies that the optimization of compilers does not work well by compiling the parallelized code. In this situation, the speedup factor is limited compared to the sequential execution. This is because the efficiency of parallel computations is lower than that of sequential execution.

There are many reasons for this phenomenon to occur. One of the major reasons comes from global variables for the loop induction variables. For example, variable rank for MPI is usually used for a global variable to parallelize the loop (See the previous codes for the line <10>. The loop induction variables, **jstart** and **jend** are defined with the global rank variable). If the loop induction variables are global variables, code analysis by compilers will be difficult, and then the quality of optimizations by compilers deteriorates. The main reason for this is that compilers cannot find a constant number through the global variables.

As mentioned above, control variables for MPI, such as management variables for the total number of MPI and rank number, can take global variables. In this case, users should consider their performance.

Exercises

1. Make a hybrid MPI program that shows matrix-vector multiplications with a parallel computer. Evaluate the performance between pure an MPI program and a hybrid MPI program. Do you find differences in the performance? To evaluate the performance, check the number of sockets in the node, then determine the most effective way for hybrid MPI execution. Evaluate the performance according to whether it is fast or not.
2. Survey applications where hybrid MPI execution is faster than pure MPI execution.

References

1. Manual for Large-Scale Visualization System UV2000, Information Technology Center (Nagoya University, 2016), http://www.icts.nagoya-u.ac.jp/ja/sc/pdf/uv2000manual_20160311.pdf
2. P.S. Pacheco, *Parallel Programming with MPI* (Morgan Kaufmann, 1996)
3. NVIDIA NVLINK, http://www.nvidia.com/object/nvlink.html
4. ScaLAPACK—Scalable Linear Algebra PACKage, http://www.netlib.org/scalapack/
5. LAPACK–Linear Algebra PACKage, http://www.netlib.org/lapack/
6. BLAS (Basic Linear Algebra Subprograms), http://www.netlib.org/blas/

Chapter 5
Application of Techniques for High-Performance Computing

Takahiro Katagiri

Abstract This chapter gives an overview of techniques for high-performance computing with actual examples of adaptations. In addition, several techniques of speedups for communications are also shown.

5.1 Overviews of Performance Tuning

To establish performance tuning, considering the cost between gains of speedup and tuning is important. For example, if we obtain $10\times$ speedup after a 1-h tuning, then it is a great justification for tuning. But if we obtain 10% speedup after tuning for 10 days, then it is small gain for tuning. It is difficult to estimate such cost for beginners; hence, performance tuning is the final work for programmers to do. However, if the program takes 1 month to execute, then it means that the program should be speeded up.

Generally speaking, the best way is to consider algorithms during the first phase of development. Code optimization is the final phase to reduce computational amounts.

If a better algorithm cannot be adapted to your program, or your programs are already using the best algorithm possible, then it is time to apply code optimizations. The steps for code optimizations are: specifying better compiler options, modification of codes, and improvement of job allocation, and so on, to consider system aspects to establish speedup.

One of the important things for code tuning is to locate hot spots, which are the heaviest part of execution in a program. It starts to know the hot spots by measuring the execution time, or by using a tool for performance profiling. By doing this, we can know where the bottleneck parts are, which are critically affecting the performance. Some categorizations for the bottleneck parts are as follows:

- Computational Time Bottleneck: happens when computation time takes up most of the runtime.

T. Katagiri (✉)
Information Technology Center, Nagoya University, Nagoya, Japan
e-mail: katagiri@cc.nagoya-u.ac.jp

© Springer Nature Singapore Pte Ltd. 2019
M. Geshi (ed.), *The Art of High Performance Computing for Computational Science, Vol. 1*, https://doi.org/10.1007/978-981-13-6194-4_5

- Communication Time Bottleneck: happens when communication time takes up most of the runtime.
- I/O Bottleneck: happens when I/O time takes up most of the runtime.

Using computational complexity to estimate the bottleneck may not be useful in finding the real bottleneck, which usually consists of loops or functions (procedures). This is because actual performance depends on computer environments and/or execution conditions, such as problem size, etc. Hence, there may be many hot spots in unknown parts.

Hot spots also change according to the status of tuning. If the computational complexity is huge, but the problem size is small, and all data are stored in cache, then the ratio of computational time will be small. If the communication volume is small, but the number of communications is large, then the communication time will be dominant. If the volume of I/O is small, but the system has poor I/O performance, then the time of I/O will be dominant.

5.1.1 Computational Time Bottleneck

It is better to consider the following approaches for computational time bottleneck with respect to the amount of code modifications:

1. Modification of Compiler Options: Specify compiler options for code optimization, such as prefetch, software pipelining, and/or specify compiler directives, such as unrolling, tiling (blocking).
2. Modify algorithm to accept lower computation complexity.
3. Modify algorithm to accept cache optimization, thus using "blocking algorithm."
4. Adapt hand tunings that compiler cannot perform well, such as making a code with unrolling and/or apply different data structures that provide continuous accesses to obtain high performance. The adaptation of different data structures is needed for the entire modification of the program.

5.1.2 Communication Time Bottleneck

The following ways should be considered in the case of communication time bottleneck:

- Communication latency is the dominant aspect (A large number of communications): (1) Merge communications, and make one message (vectorization of communication); (2) Use redundant computations to reduce communications (increase computation complexity); (3) Hide communication time by using asynchronous communications.

- Volume of messages is the dominant aspect (Large amounts of messages per communication): (1) Use redundant computations to reduce communications (increase computation complexity); (2) Use faster communication API (such as Remote Direct Memory Access (RDMA)); (3) Hide communication time by using asynchronous communications.

5.1.3 I/O Bottleneck

The following ways should be considered in the case of I/O bottleneck:

1. Reduce amount of I/O to reduce chances of I/O.
2. Change system parameters of the OS, such as modifying I/O striping sizes. For example, increase the stripe size if the I/O consists of a big data.
3. Fast file system provided by the system is used. For example, use a staging system for input and output of files.
4. Use the faster I/O API. Each MPI processes output a file simultaneously in many implementations. Utilize fast I/O API, such as Parallel I/O, or MPI-IO, to output one file to integrate multiple files by each MPI process.

5.2 Performance Profiling

To locate hot spots in your program, using a performance profiler is one of the simplest ways. Available performance profilers are different in each system. Dedicated performance profilers are usually installed in supercomputers.

The contents of profiling depend on the profilers. Usually, we can obtain the following contents of performance.

- Inner node performance: Ratios of execution time for each function (procedure) to the whole runtime, GFLOPS ratios, cache hit ratios, efficiency of thread parallelization (load balancing), ratio of I/O time, etc.
- Internode performance: Communication pattern of MPI, size of communications, number of communications. Many of these contents can be seen by a Graphical User Interface (GUI). If a GUI is not available, it may be difficult to understand the communication pattern.

5.3 Examples of Loop Transformations and Their Effects

In this section, we give an example of code tuning to adapt loop split and loop fusion (loop collapse) for an application.

Loop split and loop fusion (loop collapse) are adapted to the hot spots of Seism3D [1], which is a simulation code of seismic wave analysis developed by Professor Furumura at the University of Tokyo. An overview of the hot spots in Seism3D was explained in the example in Sect. 1.3.6.

We adapt the following seven kinds of loop transformations: **#1**: Original three nested loops (Baseline); **#2**: Loop split for I-loop; **#3**: Loop split for J-loop; **#4**: Loop split for K-loop; **#5**: Loop fusion (Loop Collapse) for code of #2 (Making two nested loop); **#6**: Loop fusion (Loop Collapse) for code of #1 (Making a one nested loop); **#7**: Loop fusion (Loop Collapse) for code of #1 (Making a two-nested loop).

The above seven kinds of codes are evaluated with the FX10 supercomputer system [CPU: Sparc64 IVfx (1.848 GHz)], which is installed in the Information Technology Center, The University of Tokyo, with one node (16 threads). Figure 5.1 shows the result of the evaluation.

Figure 5.1 shows that implementation #4 is the fastest with a factor of $1.5\times$ speedup to implementation #1 (original). This indicates that loop transformation is effective for this example. In the implementation #4, the original three-nested loop is divided into two parts. The code is shown as follows.

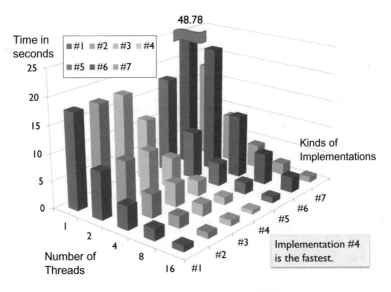

Fig. 5.1 Effect of loop transformations for Seism3D with the FX10

```
!$omp parallel do private(k,j,i,STMP1,STMP2,STMP3,STMP4,RL,
!$omp& RM,RM2,RMAXY,RMAXZ,RMAYZ,RLTHETA,QG)
DO K = 1, NZ
  DO J = 1, NY
    DO I = 1, NX
      RL = LAM (I,J,K); RM = RIG (I,J,K); RM2 = RM + RM;
      RLTHETA = (DXVX(I,J,K)+DYVY(I,J,K)+DZVZ(I,J,K))*RL
      QG = ABSX(I)*ABSY(J)*ABSZ(K)*Q(I,J,K)
      SXX(I,J,K)=(SXX (I,J,K)+(RLTHETA+RM2*DXVX(I,J,K))*DT)*QG
      SYY(I,J,K)=(SYY (I,J,K)+(RLTHETA+RM2*DYVY(I,J,K))*DT)*QG
      SZZ(I,J,K)=(SZZ (I,J,K)+(RLTHETA+RM2*DZVZ(I,J,K))*DT)*QG
    ENDDO
  ENDDO
ENDDO
!$omp end parallel do


!$omp parallel do private(k,j,i,STMP1,STMP2,STMP3,STMP4,RL,
!$omp& RM,RM2,RMAXY,RMAXZ,RMAYZ,RLTHETA,QG)
DO K = 1, NZ
  DO J = 1, NY
    DO I = 1, NX
      STMP1 = 1.0/RIG(I,J,K); STMP2 = 1.0/RIG(I+1,J,K);
      STMP4 = 1.0/RIG(I,J,K+1); STMP3 = STMP1 + STMP2
      RMAXY = 4.0/(STMP3+1.0/RIG(I,J+1,K)+1.0/RIG(I+1,J+1,K))
      RMAXZ = 4.0/(STMP3+STMP4+1.0/RIG(I+1,J,K+1))
      RMAYZ = 4.0/(STMP3+STMP4+1.0/RIG(I,J+1,K+1))
      QG = ABSX(I)*ABSY(J)*ABSZ(K)*Q(I,J,K)
      SXY(I,J,K)=(SXY (I,J,K)+(RMAXY*(DXVY(I,J,K)+
      DYVX(I,J,K)))*DT)*QG
```

$$SXZ(I,J,K)=(SXZ\ (I,J,K)+(RMAXZ*(DXVZ(I,J,K)+$$
$$DZVX(I,J,K)))*DT)*QG$$
$$SYZ(I,J,K)=(SYZ\ (I,J,K)+(RMAYZ*(DYVZ(I,J,K)+$$
$$DZVY(I,J,K)))*DT)*QG$$

  END DO
 END DO
END DO
!$omp end parallel do

5.4 Method for Communication Optimizations

In this section, we explain the method for communication optimization.

5.4.1 Size and Amount of Messages

Before communication optimization, we need to know the communication pattern for our programs. But, we would need to know the variations of communication time according to the size of the messages. Figure 5.2 shows a model for the communication time.

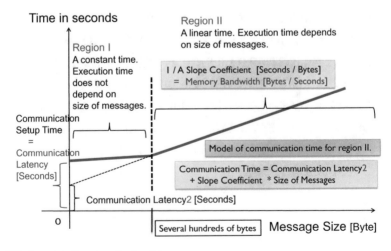

Fig. 5.2 A model of communication time

The communication pattern differs from Region I and Region II in Fig. 5.2. The way to optimize communications is also different.

- If your program is in Region I, then communication latency dominates the execution time. Reducing the number of communications, or/and gather communications, are one of the ways to reduce the communication time.
- If your program is in Region II, communication time for messages dominates the execution time. Reducing the size of messages per communication, and/or increasing computation complexities to do redundant computations, are some of the ways to reduce the communication time.

We explain computations in Region I with an example, which is a reduction operation for computations of the inner product, which is implemented with MPI_Allreduce. In this example, the size of messages per communication is one (8 bytes for double precision.) With respect to 8 bytes communication, the time taken for MPI_Allreduce between with 8 bytes and several doubles, such as four doubles (32 bytes), is almost the same for many environments. Hence, it is possible to shorten the communication time by one communication of MPI_Allreduce with 32 bytes to gather 4 MPI_Allreduces with 8 bytes.

For example, there are inner products in the Conjugate Gradient (CG) Method for solving linear equations. Usually, it requires three times per iteration, hence it tends toward being a communication latency program. By modifying the algorithm to reduce communication time that provides one communication per k-iterations, it can reduce communication time with a factor of $1/k$. However, this simple approach cannot yield a converged answer. Currently, Communication Avoiding CG (CACG) is studying to establish this communication reduction for CG algorithm with respect to the accuracy of the converged answer.

5.4.2 Non-blocking Communication

One of the major approaches to reduce communication time is by reducing waiting time for communications. To establish this, reducing synchronization points is effective in many cases. To reduce the synchronization points, it requires the changing of communication modes of MPI.

Default communications tend to set to synchronization mode. For example, the sending process cannot finalize the sending until that corresponding receiving is received for the sending data, and then the sending buffer can be reused. These functions can be implemented with blocking functions for MPI. For example, MPI_Send is a blocking function.[1]

---

[1] There are many communication modes for MPI. For the default communication mode, the sending process is finished before the corresponding received calls if there is enough buffer of communications. Hence if the buffer of communications is small, sending does not follow this behavior. Thus, the sending should be held until calling the corresponding receiver.

In the chart below, "are" should be deleted before "occurred".

▸ **If process 0 has required data, then:**

Inefficient recvs are occurred in continuous send.

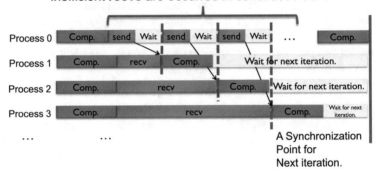

Fig. 5.3   A case in which speedup is established with a non-blocking function.

Fig. 5.3 A case in which speedup is established with a non-blocking function

On the other hand, the sending process finalizes the sending before the corresponding receiving is received for sending data for non-blocking functions of MPI. It is necessary to check that the sending buffer can be reused before proceeding to the next process. This sending allows us to reduce synchronous points. These functions can be implemented with non-blocking functions for MPI. For example, MPI_Isend is a non-blocking function.

We show a worst case for the communication with synchronization in this session. This is also the best case for using non-blocking functions. Figure 5.3 shows a case in which process #0 sends to processes #1 to #3.

In Fig. 5.3, process #0 sends data to the other processes sequentially. Hence if we use a synchronous (a blocking) sending in each sending, the other processes have waiting time in each sending. Hence, all the cost of the waiting time is increasing on an order of $p - 1$, where p is the number of processes.

With the non-blocking function to implement the sending for process #0 in Fig. 5.3, the following procedures are made to reduce communication time: (1) data sending with non-blocking function; (2) Perform required computations; (3) Check the sending. This procedure can also make computations during sending, thus hiding the sending time.

5.4.3 *Persistent Communication*

It has no meaning that **MPI_Isend** cannot send data immediately when it calls the function. However, some implementations of MPI do not provide this implementation but start sending for **MPI_Isend** when **MPI_Wait** calls. This implementation provides no effect for overlapping of communications with **MPI_Isend**.

By using **persistent communication**, it provides the effect of non-blocking communications. The effect also depends on the implementation of persistent communications of the MPI library. MPI functions for persistent communication have been supported since MPI-1. Hence, almost all environments of MPI support persistent communication. Again, it is a different problem for supporting communication overlapping and using persistent communication. It depends on the implementations for persistent communication to establish overlap between communications and computations.[2]

The procedure to implement persistent communication is as follows:

1. An initialization function to receivers is called before entering loops with communications.
2. Then, write **MPI_Start** in part of **MPI_Send**.
3. Function for synchronization points is used for the same function, such as **MPI_Wait**.

By setting the information for communication with **MPI_Send_Init**, no setting is performed in MPI_Send. Hence, the performance of **MPI_Start** is the same or better compared to MPI_Send. The restriction of persistent communication is that multiple sendings should be performed to the same receiver. Targets of persistent communication are **explicit method** based on **domain decomposition method**, or **implicit method** based on **iterative method**, for example.

The following is an example code to implement the procedure in Fig. 5.3.

---

[2]The Fujitsu PRIMEHPC FX100 system, which is installed in Information Technology Center, Nagoya University, has been supporting the overlapping between communications and computations by hardware, named **assistant core**, for dedicated hardware (CPU) for communications.

```fortran
    integer istatus(MPI_STATUS_SIZE)
    integer irequest(0:MAX_RANK_SIZE)
    ...
    if (myid .eq. 0) then
      do i=1, numprocs-1
        call MPI_SEND_INIT(a, N, MPI_DOUBLE_PRECISION, i,
            0, MPI_COMM_WORLD, irequest(i), ierr)
        enddo
      endif
    ...
    if (myid .eq. 0) then
      do i=1, numprocs-1
        call MPI_START (irequest, ierr )
      enddo
    else
      call MPI_RECV( a, N, MPI_DOUBLE,_PRECISION,
          0, i_loop, MPI_COMM_WORLD, istatus, ierr )
    endif

    Computations with a(*)

    if (myid .eq. 0) then
      do i=1, numprocs - 1
        call MPI_WAIT(irequest(i), istatus, ierr )
      enddo
    endif
```

5.5   Adaptation of Software Auto-Tuning

Recently, the cost of performance tuning is getting higher due to more complex computer architectures. As a result, the development cost to make high-performance software is also high, and this will be a problem. The performance tunings require different knowledge, hence it is very difficult to do performance tuning for non-expert users. Only experts or meisters can perform performance tuning. This also indicates that performance tuning is not engineering. Establishing methodology for performance tuning is sorted as a field of software engineering. But to obtain high-performance, execution speed for the products is required, while conventional software engineering treats productivity, and tends to not focus on the execution speed of the products. Hence, some technologies from software engineering are not adaptable for performance tuning.

Establishing high-performance computing with software engineering is sorted as **Software Performance Engineering** (SPE). This is where one applies technologies for development to high-performance software.

To automate development of high-performance software, **Software Auto-tuning** (**AT**) [2–10] has been studied since around 2000. In particular, AT researches are very active in Japan. Many of the researches deal with AT theory and performance modeling for AT [3–5], dedicated AT languages [6–10], and adaptation of applications.

Code optimization technologies shown in this book, such as **loop unrolling** and **loop transformations** (loop split and loop fusion (loop collapse)), are implemented with AT languages, named **ppOpen-AT** [8–10]. Hence, users can easily apply AT with directives provided by ppOpenAT for specifying loop unrolling, for example. By specifying these directives, codes of candidate implementations and auto-tuner are automatically generated by the preprocessor of ppOpen-AT. In addition, the generated AT codes are automatically added to the original program. Hence by adapting ppOpen-AT, the cost of development for AT software can be dramatically reduced.

5.6   A Short Note of Using Automatic Parallelization Compilers

In the final part of this chapter, the author mentions a short note to automatic parallelization compilers. Reader should understand the fact that automatic parallelization compilers never modify algorithms or programs for parallelization. This implies that the original code should have parallelism embedded in the algorithm and description in code level. Hence, users cannot obtain parallel code from algorithmic sequential code by adapting automatic parallelization compilers.

The author believes that automatic parallelization compilers strongly require a *well-described* code. If an algorithm has parallelism, but the description of the sequential code is unfit for parallelization, then users must obtain a sequential code via automatic parallelization compiler. Parallelization by automatic parallelization compilers also depends on computer language specification. For example, loop structures, such as starting and ending values for each loop are clearly stated or not, may affect the parallelization. Whether arguments are passed by value or by reference can also affect the parallelization, since a loop that is inside the function (procedures) is using the arguments, the compiler may find *false* dependency for values via the arguments. If compilers decide to have dependency for the values to make correct codes, then the generated codes are sequentialized. As a result, even if we write obvious parallel codes inside a function (a procedure), the compiler cannot generate parallel code.[3]

The above examples indicate that users should learn how to parallelize their code. This also indicates that the cost of programming between automatic parallel compiler and OpenMP or MPI will be almost the same. Hence it should be mentioned that using an automatic parallelization, compiler will not always reduce the cost of parallelization.

Exercises

1. Survey the performance profilers in your environment. Then use them.
2. Implement sample programs shown in this chapter with available parallel computers. Evaluate the performance between non-blocking sending [MPI_Isend function (procedure)] and blocking sending [MPI_Send function (procedure)], then find the effective size of messages for MPI_Isend by varying it from 1 to a reasonable upper limit.
3. Implement the program for #2 with functions (procedures) of persistent communication. Compare the execution time between persistent communication and non-blocking communication.
4. Adapt loop split, loop fusion (loop collapse) to your code. Evaluate the performance code with OpenMP by varying the number of threads.
5. Survey the tools and computer languages for Software Auto-tuning (AT).
6. Download codes of ppOpen-AT from the homepage of ppOpen-HPC project [11]. Install the ppOpen-AT to your computer. Evaluate the effect of AT by using the attached sample programs.

References

1. T. Furumura, L. Chen, Parallel Comput. **31**, 149 (2005)
2. D.A. Patterson, J.L. Hennessy, in *Computer Organization and Design MIPS Edition, Fifth Edition: The Hardware/Software Interface*. The Morgan Kaufmann Series in Computer Architecture and Design (2013)

---

[3]In this case, users should write a directive or describe a compiler option to know nondependency for the argument values to compilers. The readers can survey these directives and options in their compilers before using it.

3. T. Katagiri, K. Kise, H. Honda, T. Yuba, in *The Fifth International Symposium on High Performance Computing (ISHPC-V)*. Springer Lecture Notes in Computer Science, vol. 2858 (2003), p. 146
4. T. Tanaka, R. Otsuka, A. Fujii, T. Katagiri, T. Imamura, Scientific Program. **22**, 299 (2014). IOS Press
5. R. Murata, J. Irie, A. Fujii, T. Tanaka, T. Katagiri, in *Proceedings of Embedded Multicore/Many-core Systems-on-Chip (MCSoC)* (2015), p. 203
6. T. Katagiri, K. Kise, H. Honda, T. Yuba, Parallel Comput. **32**, 92 (2010)
7. T. Katagiri, S. Ohshima, M. Matsumoto, in *Proceedings of IEEE MCSoC-2014, Special Session: Auto-Tuning for Multicore and GPU (ATMG-14)* (2014), p. 91
8. T. Katagiri, S. Ohshima, M. Matsumoto, in *Proceedings of IEEE IPDPSW 2015* (2015), p. 1221
9. T. Katagiri, S. Ohshima, M. Matsumoto, in *Proceedings of IEEE IPDPSW 2016* (2016), p. 1488
10. T. Katagiri, S. Ohshima, M. Matsumoto, in *Proceedings of IEEE IPDPSW 2017* (2017), p. 1399
11. ppOpen-HPC Project Home Page, http://ppopenhpc.cc.u-tokyo.ac.jp/ppopenhpc/

Chapter 6
Basics and Practice of Linear Algebra Calculation Library BLAS and LAPACK

Maho Nakata

Abstract In this chapter, we explain the basic architecture and use of the linear algebra calculation libraries called BLAS and LAPACK. BLAS and LAPACK libraries are for carrying out vector and matrix operations on computers. They are used by many programs, and their implementations are optimized according to the computer they are run on. These libraries should be used whenever possible for linear algebra operations. This is because algorithms based directly on mathematical theorems in textbooks may be inefficient and their results may not have sufficient accuracy in practice. Moreover, programming such algorithms are bothersome. However, performance may suffer if you use a non-optimized library. In fact, the difference in performance between a non-optimized and optimized one is likely very large, so you should choose the fastest one for your computer. The availability of optimized BLAS and LAPACK libraries have improved remarkably. For example, they are now included in Linux distributions such as Ubuntu. In this chapter, we will refer to the libraries for Ubuntu 16.04 so that readers can easily try them out for themselves. Unfortunately, we will not mention GPU implementations on account of lack of space. However, the basic ideas are the same as presented in this chapter; therefore, we believe that readers will easily be able to utilize them as well.

6.1 Importance of Linear Algebra

Linear algebra has many applications, and it is a very important building block for many things. The basics are typically learned in the first year of university, and they are as abstract as they are fundamental. The readers may be quick to understand though that learning while considering applications has its advantages. In this section, we explain the historical aspects and modern usages of linear algebra so that readers will be able to imagine the correspondences of vectors and matrices in real applications.

M. Nakata (✉)
RIKEN Advanced Center for Computing and Communication, Wako, Japan
e-mail: maho@riken.jp

© Springer Nature Singapore Pte Ltd. 2019
M. Geshi (ed.), *The Art of High Performance Computing for Computational Science, Vol. 1*, https://doi.org/10.1007/978-981-13-6194-4_6

6.1.1 Historical Things, Especially in the First Time Gaussian Elimination was Used in Ancient China

Linear algebra has existed for a long time. The first descriptions are on fragments of papyrus found in Egypt. Here, we would like to illustrate the first use of the Gaussian elimination method in ancient China. One record of it is written in a book, titled "The Nine Chapters on the Mathematical Art" [1], that was in circulation from the first century BC to the second century AD. Chapter 8 of this book, "Rectangular Arrays", contains 18 problems and solutions to simultaneous linear equations obtained by using what we call today Gaussian elimination. The first problem, which outlines solution method, is posed as follows:

- 今有上禾三秉, 中禾二秉, 下禾一秉, 實三十九斗；上禾二秉, 中禾三秉, 下禾一秉, 實三十四斗；上禾一秉, 中禾二秉, 下禾三秉, 實二十六斗. 問上, 中, 下禾實一秉各幾何.

- 答曰:上禾一秉, 九斗, 四分斗之一, 中禾一秉, 四斗, 四分斗之一, 下禾一秉, 二斗, 四分斗之三.

- 方程術曰, 置上禾三秉, 中禾二秉, 下禾一秉, 實三十九斗, 於右方. 中, 左禾列如右方. 以右行上禾遍乘中行而以直除. 又乘其次, 亦以直除. 然以中行中禾不盡者遍乘左行而以直除. 左方下禾不盡者, 上為法, 下為實. 實即下禾之實. 求中禾, 以法乘中行下實, 而除下禾之實. 餘如中禾秉數而一, 即中禾之實. 求上禾亦以法乘右行下實, 而除下禾, 中禾之實. 餘如上禾秉數而一, 即上禾之實. 實皆如法, 各得一斗.

Here is a partial English translation:

- Q: There are three bundles of fine millet, two bundles of medium-quality millet, one bundle of low-quality millet, filling 39 buckets in total. Next, there are two bundles of fine millet, three bundles of medium-quality millet, and one bundle of low-quality millet, 34 buckets in total. Finally, there is one bundle of fine millet, two bundles of medium-quality millet, and three bundles of low-quality millet, 26 buckets in total. How many buckets are there of one bundle of high-quality, of medium-quality, and of low-quality millet?
- A: fine $9\frac{1}{4}$ buckets, medium $4\frac{1}{4}$ buckets, low $2\frac{3}{4}$ buckets.
- Put the three bundles of fine-quality, two bundles of medium-quality, and one bundle of poor-quality millet and 39 buckets in a column on the right. Put the second set in the middle, and put the third set on the left, respectively, like follows:

L	M	R
1	2	3
2	3	2
3	1	1
26	34	39

Multiply the middle by the buckets of fine-quality mille of the right (therefore, we multiply the middle column by three); then, subtract the right from it until the bundles of fine millet become zero.

1	2	3		1	6	3		1	3	3		1	0	3
2	3	2		2	9	2		2	7	2		2	5	2
3	1	1	→	3	3	1	→	3	2	1	→	3	1	1
26	34	39		26	102	39		26	63	39		26	24	39

The left column is multiplied and subtracted from the right column like the middle.

1	0	3		3	0	3		0	0	3
2	5	2		6	5	2		4	5	2
3	1	1	→	9	1	1	→	8	1	1
26	24	39		78	24	39		39	24	39

The left column is multiplied by five (the resultant middle entry of the middle column); then, subtract by the middle column until the middle entry of the left column is eliminated.

0	0	3		0	0	3		0	0	3
4	5	2		20	5	2		0	5	2
8	1	1	→	40	1	1	→	36	1	1
39	24	39		195	24	39		99	24	39

On the left, we have some bundles of low quality; finally, we divide the buckets on the left by them, and we have the answer for the low-quality buckets. Now, let us obtain medium… (the rest is omitted).

When we express this problem in terms of modern arithmetic, it becomes as follows:

- Q: Let x be the number of buckets of high-quality millet, y the number of buckets of medium-quality millet, and z the number of buckets of low-quality millet. Solve the following simultaneous equations for x, y, and z.

$$\begin{cases} 3x + 2y + z = 39 \cdots (R) \\ 2x + 3y + z = 34 \cdots (M) \\ x + 2y + 3z = 26 \cdots (L) \end{cases}$$

(置上禾三秉, 中禾二秉, 下禾一秉, 實三十九斗於右方, 中, 左禾列如右方.)

- Multiply (M) by 3 and (R) by 2; then subtract (M) from (R). Next, multiply (L) by 3; then subtract (R) from (L). (以右行上禾遍乘中行, 而以直除. 又乘其次, 亦以直除.)

$$3(2x + 3y + z) = 34 \times 3$$
$$\underline{2(3x + 2y + z) = 39 \times 2}$$
$$5y + z = 24 \cdots (M)$$

$$3(x + 2y + 3z) = 26 \times 3$$
$$\underline{3x + 2y + z = 39}$$
$$4y + 8z = 39 \cdots (L)$$

- Then, we have following

$$
\begin{cases}
3x + 2y + z = 39 \cdots (R) \\
\qquad 5y + z = 24 \cdots (M) \\
20y + 40z = 195 \cdots (L)
\end{cases}
$$

- Next, compute (L) – (M) × 4 (然以中行中禾不盡者 (=5) 遍乘左行而以直除) and obtain z as follows (左方下禾不盡者, 上為法, 下為實. 實即下禾之實)

$$36z = 99$$
$$4z = 11$$
$$z = 2\tfrac{3}{4}$$

... (the rest is omitted).

The book was intended for administrative purposes, and the method and linear algebra was not developed any further in ancient China. In fact, it was forgotten for a considerable period, until the middle of the Qing dynasty (1644–1912). Carl Friedrich Gauss rediscovered it at the beginning of the nineteenth century.[1] Nowadays, it is little mentioned that the ancient Chinese knew the method, but that it was forgotten to history.

6.1.2 Examples of Use in Modern Times

Linear algebra is used everywhere today. Let us list some examples.

- Modern computer chips: The principles underlying computer chips are those of quantum mechanics. Mathematically, quantum mechanics is based on infinite-dimensional linear algebra called Hilbert space theory. The wavefunctions of quantum mechanics are vectors, while physical quantities (observables) are linear operators (matrices). For example, the total energy of the system corresponds to the eigenvalue of the operator (matrix) called the Hamiltonian, and it is used to solve the eigenvalue problem of a complex-valued Hermitian matrix.
- 3D CGs: To show the appearance of a three-dimensional object moving on a display, a computer performs a coordinate conversion by linear algebra tens of thousands of times per second.

---

[1] Although Carl Friedrich Gauss is sometimes credited with the rediscovery, Isaac Newton, a 100 years earlier, wrote that textbooks of the day lacked a method for solving simultaneous equations and proceeded to publish one that became well circulated.

- Structural analysis: Structural analysis is necessary for the design of buildings, and it entails solving eigenvalue problems.
- Linear programming: Linear programming is used to maximize, for example, the efficiency of transportation resources and for portfolio optimization in finance. The simplex algorithm involves solving linear simultaneous linear equations.
- Machine learning, deep learning: The basic algorithm is a mass of linear algebra. Various linear algebra operations including singular value decomposition and matrix–matrix multiplication is used.
- Least squares method: It approximates a specific function with respect to numerical values obtained from experiments, etc. It uses linear algebra to make an approximation that minimizes the sum of squares of residuals (a residual is the difference between an observed value and the fitted value provided by a mathematical model).

Linear algebra appears in such various fields that it is a rare situation that you can not talk of its use. Since its range of application is wide, it has a high degree of abstraction. Various textbooks, commentaries, web documents, and classroom videos on the internet has been published on it, so please do your best to learn from them.

6.2 Importance of Using Pre-developed Libraries

Compared to what they were 10 years ago, the basic libraries have improved significantly in terms of availability and quality. Here, we describe the importance of using such libraries; although it may seem to be lazy and easy going, it is better to first search for existing libraries rather than to reinvent the wheel.

It is a complicated and troublesome task to organize a library, but the alternative has a number of drawbacks. Algorithmic implementations of mathematical theorems straight out of textbooks may be slow and may have a large numerical error; most importantly, they may not fully utilize the available CPU power.

6.2.1 Handling Real Numbers in a Computer: Floating-Point Numbers

Computers naturally handle only bit sequences of zeroes and ones. This essentially means that they can easily operate on integers, but not real numbers. As a way of representing real numbers, they use special numbers called "floating-point numbers." Almost, all real number calculations on computers are done using floating-point numbers. The formats of floating-point numbers are defined in IEEE 754-2008 [2]. These formats are characterized by precision, i.e., (i) binary32, (ii) binary64, and (iii) binary128, and are customarily called single precision, double precision, and quadruple precision. As the names imply, the binary32 format occupies 32 bits, the binary64 format 64 bits, and the binary128 format 128 bits, respectively. A figure illustrating the binary64 floating-point number format is shown in Fig. 6.1.

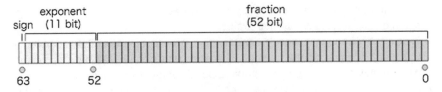

Fig. 6.1 Figure illustrating a binary64 floating-point number

Here, binary64, or double precision, the format includes a sign of 1 bit, an exponent part with 11 bits, and a fraction part with 52 bits.[2]

$$(-1)^{\text{sign}} \times 1.\overbrace{01010010100001\cdots 11}^{52\text{bit}} \times 2^{\text{exponent}}.$$

Actually, the exponent part is biased by 1023 or 1023 is subtracted from it to omit the sign. Therefore, the range of the exponent is -1022 to 1023. This type of number has approximately 16 decimal significant digits, computed as $\log_{10} 2^{53} = 15.96$.

Since floating-point numbers are not "real" numbers, any calculation with them that is intended to represent a calculation involving real numbers they may have errors. Furthermore, there are limitations in the number of numbers that can be handled. We show an example in which the associative law $a + (b + c) = (a + b) + c$ is not satisfied. In real numbers, the following equality holds:

$$1.0 + (-1.0 + 1.0 \times 10^{-20}) = (1.0 - 1.0) + 1.0 \times 10^{-20} = 1.0 \times 10^{-20}.$$

But with double-precision floating-point arithmetic, we get

$$1.0 + (-1.0 + 1.0 \times 10^{-20}) = 0.$$

We also show an example in which the distributive rule also does not hold.

$$(1.62 - 0.983) \times 3.15 = 2.00655.$$

For the sake of simplicity, let us assume there are three decimal significant digits in a floating-point number. In this example, a calculation without the distributive rule yields,

$$(1.62 - 0.983) \times 3.15 = 2.01.$$

On the other hand, when we calculate with the distributive rule, we get

$$1.65 \times 3.15 - 0.983 \times 3.15 = 5.10 - 3.10 = 2.00.$$

---

[2]There once was a time when the format varied from one manufacturer or vendor to the other; when data ceased being compatible or when a computer was replaced, it was necessary to change the program as well.

In this example, the same mathematical calculations gave different results by the floating-point calculation depending on the order of calculation. Note that, the commutative law holds for floating-point number arithmetic. Anyway, normally, we do not have to worry too much save that when we treat floating-point numbers, we should be aware that numerical errors may occur. That is, the same calculation may give different results.

For example, using BLAS optimized for recent multicore CPUs, the same matrix–matrix multiplication performed twice might give slightly different results. The reason is that in the usual optimized implementation, matrices are divided into sub-matrices (see Sect. 6.4.3). Operations on sub-matrices can be performed independently (thus, we can utilize multicore CPUs), and then the sub-matrices are added to obtain the final answer. Because of this independence, the order of addition of sub-matrices may be different. Therefore, the violation of the associative law occurs and it may give different results.

6.2.2   Example in Which a Textbook-Like Implementation is Inefficient

Let A be an $n \times n$ matrix, and let x and y be n-dimensional vectors. Let us consider solving linear simultaneous linear equation by using Cramer's rule:

$$Ax = y.$$

Cramer's rule states that when $x = (x_1, \ldots, x_n)^t$,

$$y_i = \frac{\det(A_i)}{\det(A)}$$

where A_i is the matrix formed by replacing the ith column of A by the column vector y. In principle, it should be able to solve the linear simultaneous linear equations when a solution exists. In order to obtain one determinant, we need $(n - 1) \times n! + n! - 1$ arithmetic operations. This becomes more and more computationally expensive as n gets bigger; for example, if $n = 100$, $n! = 9.3 \times 10^{157}$ operations are required. Even the fastest computer on the planet can calculate only "10^{18} times" in a second; such a calculation is not realistic at all.

Then, how about calculating the inverse matrix?:

$$x = A^{-1}y.$$

In principle, the matrix inverse can be obtained by applying Cramer's rule. However, matrix inversions should be avoided because they can easily cause numerical errors (see the discussion on p. 260 of [3]).

Normally, in order to solve such a linear equation by a computer, LU decomposition is used. When this is done, the number of calculations is roughly $n^3/3$, i.e., much faster than inverting matrices by using Cramer's rule. Moreover, a solution can be calculated with less numerical error.

6.2.3 Example of Fast Optimized Implementation

Now, let us look at the difference between a slow (reference) implementation and a fast (optimized) one. In particular, the performance of matrix–matrix multiplication dramatically changes. Let's start by looking at the results of an experiment on an Intel Xeon E5-2603 @ 1.80 GHz 4 core running Ubuntu 16.04. The readers may want to set up a machine running Ubuntu 16.04 to reproduce the results.

The software for comparison uses GNU Octave [4], which is a high-level interpreter language primarily for numerical computations, especially of linear and nonlinear problems. GNU Octave can easily handle matrices and vectors, and it uses BLAS and LAPACK internally for the calculation. Thus, it is easy to use it for a comparison. GNU Octave is a clone of MATLAB [5], which is the standard in this field. We used GNU Octave, instead, because it is free and has the added flexibility of being able to change libraries dynamically.

Now, let us compare! First, install the relevant software; then change/verify library linked against; finally, perform the matrix–matrix multiplication benchmark of $C = AB$, where A, B, and C are $n \times n$ matrices. The resulting Gflops is the performance value. In this case, a larger Gflops value is faster.

```
$ sudo apt install libopenblas-base libblas3 octave
...
$ sudo update-alternatives --config libblas.so.3
There are two choices for the alternative libblas.so.3 (providing /usr/lib/
    libblas.so.3).

  Selection    Path                                      Priority   Status
------------------------------------------------------------
* 0            /usr/lib/openblas-base/libblas.so.3        40         auto mode
  1            /usr/lib/libblas/libblas.so.3              10         manual mode
  2            /usr/lib/openblas-base/libblas.so.3        40         manual mode
<--- select 1
$ octave
...snip...
octave:1> n=4000; A=rand(n); B=rand(n);
octave:2> tic(); C=A*B; t=toc(); GFLOPS=2*n^3/t*1e-9
GFLOPS =    1.2861
...snip...
$ sudo update-alternatives --config libblas.so.3
<--- select 0
$ octave
octave:1> n=4000; A=rand(n); B=rand(n);
octave:2> tic(); C=A*B; t=toc(); GFLOPS=2*n^3/t*1e-9
GFLOPS =   94.835
octave:3>
```

In the first measurement, we used the reference BLAS (this library contains only textbook implementations), and the matrix–matrix operation ran at 1.28 Gflops (flops means how many operations per second can be done, G is 10^9, so in this case, it was able to calculate about 1.28×10^9 times in a second).

In the next measurement, we used OpenBLAS; it ran at 94.8 Gflops (meaning that we could calculate 94.8×10^9 times per second). This is 74 times faster than the reference! This is a surprising result, where the same computer running an optimized routine yielded a dramatic performance increase over running a non-optimized one. You can now see why you should use an optimized library instead of a slow one.

6.3   Introduction to BLAS and LAPACK

6.3.1   What is BLAS?

First, we let us describe the BLAS (Basic Linear Algebra Subprograms) library. As its name implies, it is a basic linear algebra subprogram and provides various routines written in FORTRAN 77 for performing matrix–matrix multiplication, matrix–vector multiplication, the inner product of two vectors, constant addition, etc. BLAS serves as a set of building blocks to do more complex things.

In addition, BLAS is provided with a reference implementation (usually called reference BLAS). The reference implementation is written very beautifully and runs on almost all computers from Raspberry Pi to supercomputers. The program itself is well documented. Although it is a reference implementation and its beauty is frequently overlooked, these aspects have made BLAS the world standard.

There are three types of BLAS, Level 1, Level 2, and Level 3. Level 1 contains vector–vector or vector operation routines and other miscellaneous routines, for example, for vector addition (DAXPY),

$$y \leftarrow \alpha x + y, \tag{6.1}$$

and computing inner products (DDOT),

$$dot \leftarrow x^T y. \tag{6.2}$$

Level 1 has 15 different operations as well as 4 combinations by precision and number type of operations, i.e., single precision, double precision, complex single precision, and complex double precision.

Level 2 is a group of matrix–vector operation routines. It contains 25 different operations, including the ones for computing the matrix–vector product (DGEMV),

$$y \leftarrow \alpha A x + \beta y, \tag{6.3}$$

and for solving simultaneous linear equations in upper triangular matrix form (DTRSV),

$$x \leftarrow A^{-1} b, \tag{6.4}$$

etc. It too has four different combinations by precision and number type, i.e., single precision, double precision, complex single precision, and complex double precision. Level 3 is a group of matrix–matrix operation routines. It contains nine different operations, including the ones for computing matrix–matrix products (DGEMM),

$$C \leftarrow \alpha A B + \beta C \qquad (6.5)$$

and (DSYRK),

$$C \leftarrow \alpha A A^T + \beta C, \qquad (6.6)$$

and for solving simultaneous linear equations in upper triangular matrix form (DTRSM),

$$B \leftarrow \alpha A^{-1} B. \qquad (6.7)$$

It also has four combinations of precision (single/double) and number type (real/-complex). If you want to see what routines are there, please refer to the BLAS Quick Reference Card [6].

6.3.2 What is LAPACK?

LAPACK (Linear Algebra PACKage) a linear algebra package that, using BLAS as a building block, solves simultaneous linear equations, least squares, eigenvalue problems, and singular value problems, which are more advanced problems of linear algebra. Also, it can decompose the matrices that appear in these problems (through LU, Cholesky, QR, singular value, Schur, and generalized Schur decomposition). It also provides a variety of subtask routines such as for estimating condition numbers and for inverting matrices. LAPACK developers are dedicated to quality assurance, making it a very reliable library. Unlike BLAS which uses Fortran 77, LAPACK uses Fortran 90 (since version 3.2). Since version 3.3, an API for use with C has been provided, and the package has been made thread safe.[3] The latest version 3.7.1 came out on June 25, 2017. LAPACK consists of approximately 2000 routines (about 500 routines for each of single precision, double precision, complex single precision, complex double precision)

It runs on a variety of CPUs and operating systems, ranging from personal computers to supercomputers, and there are 150 million visits on the website. It is a wonderful library loved by people around the world and it can be said to be a treasure of mankind.

---

[3]In multicore environments, programs run in parallel in a light process called "threads." Because different threads can access the same memory area at the same time. It may induce conflicts. In LAPACK 3.3, all the routines are now thread safe by removing such private variables.

6.4   Exercises on Using BLAS and LAPACK with C/C++

In going through the example in Sect. 6.2.3, readers have already set up Ubuntu 16.04. In this section, we use BLAS and LAPACK. The reader should first install the relevant software packages. Open the terminal, and enter as follows:

```
$ sudo apt install libblas-dev liblapack-dev liblapacke-dev octave gfortran
    g++.
```

6.4.1   BLAS Exercise: Matrix–Matrix Product DGEMM *from* C++

Let us use the matrix–matrix product DGEMM routine of BLAS library. More specifically, let's write a program to calculate

$$C \leftarrow \alpha AB + \beta C$$

where

$$A = \begin{pmatrix} 1 & 8 & 3 \\ 2 & 10 & 8 \\ 9 & -5 & -1 \end{pmatrix}$$

$$B = \begin{pmatrix} 9 & 8 & 3 \\ 3 & 11 & 2.3 \\ -8 & 6 & 1 \end{pmatrix}$$

$$C = \begin{pmatrix} 3 & 3 & 1.2 \\ 8 & 4 & 8 \\ 6 & 1 & -2 \end{pmatrix}$$

and $\alpha = 3$, $\beta = -2$. The answer is following

$$\begin{pmatrix} 21 & 336 & 70.8 \\ -64 & 514 & 95 \\ 210 & 31 & 47.5 \end{pmatrix}.$$

Input List 6.1 by using an editor like emacs or vim, and type on the terminal as follows:

```
$ g++ dgemm_demo.cpp -o dgemm_demo -lblas.
```

If there is no message, the compilation is successful. Note that including "cblas.h" and specifying that the matrix order as column major, which will be explained in the next section, is important.

Run the program on the terminal as follows:

```
$ ./dgemm_demo
# dgemm demo...
A =[ [ 1.00e+00,  8.00e+00,  3.00e+00];\
     [ 2.00e+00,  1.00e+01,  8.00e+00];\
     [ 9.00e+00, -5.00e+00, -1.00e+00] ]
B =[ [ 9.00e+00,  8.00e+00,  3.00e+00];\
     [ 3.00e+00,  1.10e+01,  2.30e+00];\
     [ -8.00e+00,  6.00e+00,  1.00e+00] ]
C =[ [ 3.00e+00,  3.00e+00,  1.20e+00];\
     [ 8.00e+00,  4.00e+00,  8.00e+00];\
     [ 6.00e+00,  1.00e+00, -2.00e+00] ]
alpha = 3.000e+00
beta  = -2.000e+00
ans=[ [ 2.10e+01,  3.36e+02,  7.08e+01];\
      [ -6.40e+01,  5.14e+02,  9.50e+01];\
      [ 2.10e+02,  3.10e+01,  4.75e+01] ]
#check by Matlab/Octave by:
alpha * A * B + beta * C
```

You can check the answer by connecting to Octave or copying and pasting the result to Octave, as follows:

List 6.1 Sample of matrix–matrix product DGEMM in C++. Filename should be "dgemm demo.cpp".

```cpp
// dgemm test public domain
#include <stdio.h>
#include <cblas.h>

//Matlab/Octave format
void printmat(int N, int M, double *A, int LDA) {
  double mtmp;
  printf("[ ");
  for (int i = 0; i < N; i++) {
    printf("[ ");
    for (int j = 0; j < M; j++) {
      mtmp = A[i + j * LDA];
      printf("%5.2e", mtmp);
      if (j < M - 1) printf(", ");
    } if (i < N - 1) printf("]; ");
    else printf("] ");
  } printf("]");
}
int main()
{
  int n = 3; double alpha, beta;
  double *A = new double[n*n];
  double *B = new double[n*n];
  double *C = new double[n*n];

  A[0+0*n]=1; A[0+1*n]= 8; A[0+2*n]= 3;
  A[1+0*n]=2; A[1+1*n]=10; A[1+2*n]= 8;
  A[2+0*n]=9; A[2+1*n]=-5; A[2+2*n]=-1;

  B[0+0*n]= 9; B[0+1*n]= 8; B[0+2*n]=3;
  B[1+0*n]= 3; B[1+1*n]=11; B[1+2*n]=2.3;
  B[2+0*n]=-8; B[2+1*n]= 6; B[2+2*n]=1;

  C[0+0*n]=3; C[0+1*n]=3; C[0+2*n]=1.2;
  C[1+0*n]=8; C[1+1*n]=4; C[1+2*n]=8;
  C[2+0*n]=6; C[2+1*n]=1; C[2+2*n]=-2;

  printf("# dgemm demo...\n");
  printf("A =");printmat(n,n,A,n);printf("\n");
  printf("B =");printmat(n,n,B,n);printf("\n");
```

```
printf("C =");printmat(n,n,C,n);printf("\n");
alpha = 3.0; beta = -2.0;
CBLAS_LAYOUT Layout = CblasColMajor;
CBLAS_TRANSPOSE trans = CblasNoTrans;
cblas_dgemm(Layout, trans, trans, n, n, n, alpha, A, n, B, n, beta, C, n);
printf("alpha = %5.3e\n", alpha);
printf("beta  = %5.3e\n", beta);
printf("ans="); printmat(n,n,C,n);
printf("\n");
printf("#check by Matlab/Octave by:\n");
printf("alpha * A * B + beta * C\n");
delete[]C; delete[]B; delete[]A;
}
```

```
$ ./dgemm_demo | octave
A =

   1    8    3
   2   10    8
   9   -5   -1

B =

    9.0000    8.0000    3.0000
    3.0000   11.0000    2.3000
   -8.0000    6.0000    1.0000

C =

    3.0000    3.0000    1.2000
    8.0000    4.0000    8.0000
    6.0000    1.0000   -2.0000

alpha =  3
beta = -2
ans =

   21.000   336.000    70.800
  -64.000   514.000    95.000
  210.000    31.000    47.500

ans =

   21.000   336.000    70.800
  -64.000   514.000    95.000
  210.000    31.000    47.500
```

If the last two matrices are the same, the answer is correct. Congratulations! If the two outputs are different, please look for differences between your program and ours and correct them. Then compile and run the program again.

The following part of the program

```
CBLAS_LAYOUT Layout = CblasColMajor;
CBLAS_TRANSPOSE trans = CblasNoTrans;
cblas_dgemm(Layout, trans, trans, n, n, n, alpha, A, n, B, n, beta, C, n);
```

is to calculate the matrix–matrix product. "CBLAS_LAYOUT" specifies how the program stores the matrix in memory. It depends on whether it is row major or column major. In this case, we choose the column major way. Please refer to following code snippet to see how to do so in practice.

```
A[0+0*n]=1;  A[0+1*n]= 8;  A[0+2*n]= 3;
A[1+0*n]=2;  A[1+1*n]=10;  A[1+2*n]= 8;
A[2+0*n]=9;  A[2+1*n]=-5;  A[2+2*n]=-1;
```

Readers also should refer to Figs. 6.2 and 6.3. Note that CBLAS_LAYOUT is row major in C/C++, but the default (column major) in FORTRAN. Since BLAS was originally written in FORTRAN, there are minor differences such as this.

"CBLAS_TRANSPOSE" specifies whether or not to perform transposition of a matrix. In this case, "CblasNoTrans" is specified because we did not choose to perform a transpose operation.

"cblas_dgemm" actually performs the matrix–matrix multiplication; one might imagine that n specifies the dimension of the matrix, but in fact, n is specified by six times. The first three are specifying the size of matrices, but the last three parameters specify the leading dimension of each matrix. This is used to treat the sub-matrices of a matrix (we will discuss this in Sect. 6.4.3).

6.4.2 LAPACK Exercise: Determining the Eigenvectors and Eigenvalues of Real Symmetric Matrices DSYEV Called from C++

Next, let us try DSYEV, which diagonalizes a real symmetric matrix, from C++ of LAPACK.

Let us find the eigenvectors and eigenvalues of A:

$$A = \begin{pmatrix} 1 & 2 & 3 \\ 2 & 5 & 4 \\ 3 & 4 & 6 \end{pmatrix}.$$

The eigenvectors are v_1, v_2, v_3:

$$v_1 = (-0.914357, 0.216411, 0.342225)$$
$$v_2 = (0.040122, -0.792606, 0.608413)$$
$$v_3 = (0.402916, 0.570037, 0.716042)$$

and the eigenvalues are:

$$-0.40973, 1.57715, 10.83258.$$

Let us reproduce these results by using the LAPACK routine. Input the List 6.2 using an editor like emacs or vim, and type the following to compile the program.

```
$ g++ eigenvalue_demo.cpp -o eigenvalue_demo -lblas -llapack -llpacke
```

If no messages are returned, the compilation was successful. Next, run this program on the terminal. The output of the program should look like

```
$ ./dsyev_demo
A =[ [ 1.000000000000000e+00, 2.000000000000000e+00, 3.000000000000000e
     +00];
[ 2.000000000000000e+00, 5.000000000000000e+00, 4.000000000000000e+00];
[ 3.000000000000000e+00, 4.000000000000000e+00, 6.000000000000000e+00] ]
#eigenvalues
w =[ [ -4.097260154436277e-01]; [ 1.577148955239774e+00]; [
     1.083257706020385e+01] ]
#eigenvecs
U =[ [ -9.143569741239890e-01, 2.164105806969943e-01, 3.422247571892085e
     -01];
[ 4.012194019687459e-02, -7.926056565391025e-01, 6.084131023712891e-01];
[ 4.029163111437181e-01, 5.700374845434359e-01, 7.160416974099556e-01] ]
#Check Matlab/Octave by:
eig(A)
U'*A*U.
```

Indeed, the eigenvalues and eigenvectors are obtained. You can check the answer by connecting to Octave or copying and pasting the result to Octave as follows:

```
$ ./dsyev_demo    | octave
A =

   1    2    3
   2    5    4
   3    4    6

w =

   -0.40973
    1.57715
   10.83258

U =

   -0.914357    0.216411    0.342225
    0.040122   -0.792606    0.608413
    0.402916    0.570037    0.716042

ans =

   -0.40973
    1.57715
   10.83258

ans =

   -4.0973e-01    1.3878e-17    8.6042e-16
    1.6653e-16    1.5771e+00   -4.4409e-16
    4.4409e-16   -8.8818e-16    1.0833e+01
```

Here, the first "ans" is the list of eigenvalues, which are as we stated. The second "ans" is the result of diagonalizing the matrix A using the eigenvectors $([v_1 v_2 v_3]^t A [v_1 v_2 v_3])$. The diagonal elements of this diagonal matrix (to within a very small numerical error) correspond to the eigenvalues.

List 6.2 Sample of DSYEV in C++ finding eigenvalues and eigenvectors. Filename should be "eigenvalue_demo.cpp"

```cpp
//dsyev test public domain
#include <iostream>
#include <stdio.h>
#include <lapacke.h>

//Matlab/Octave format
void printmat(int N, int M, double *A, int LDA) {
  double mtmp;
  printf("[ ");
  for (int i = 0; i < N; i++) {
    printf("[ ");
    for (int j = 0; j < M; j++) {
      mtmp = A[i + j * LDA];
      printf("%20.15e", mtmp);
      if (j < M - 1) printf(", ");
    } if (i < N - 1) printf("]; ");
    else printf("] ");
  } printf("]");
}
int main()
{
  int n = 3;
  double *A = new double[n*n];
  double *w = new double[n];
//setting A matrix
  A[0+0*n]=1;A[0+1*n]=2;A[0+2*n]=3;
  A[1+0*n]=2;A[1+1*n]=5;A[1+2*n]=4;
  A[2+0*n]=3;A[2+1*n]=4;A[2+2*n]=6;
  printf("A ="); printmat(n, n, A, n);
  printf("\n");
  int matrix_layout = LAPACK_COL_MAJOR;
//get Eigenvalue
  LAPACKE_dsyev(matrix_layout, 'V', 'U', n, A, n, w);
//print out some results.
  printf("#eigenvalues \n"); printf("w =");
  printmat(n, 1, w, 1); printf("\n");
  printf("#eigenvecs \n"); printf("U =");
  printmat(n, n, A, n); printf("\n");
printf("#Check Matlab/Octave by:\n");
  printf("eig(A)\n");
  printf("U'*A*U\n");
  delete[]w;
  delete[]A;
}
```

Assignment of values to matrices can be easily understood. It is the same as in DGEMM. Two arrays for the eigenvalues and eigenvectors are prepared, and the matrix is diagonalized as follows.

```
LAPACKE_dsyev(matrix_layout, 'V', 'U', n, A, n, w);
```

We should use LAPACKE to call LAPACK from C/C++. Note the lack of consistency in the designation of CBLAS, matrix layout, transpose, etc. The correspondent of "CBLAS_LAYOUT" in LAPACKE is "int matrix_layout = LAPACK_COL_MAJOR;". "CBLAS_TRANSPOSE" does not exist in LAPACKE, so we need to input a character string as follows. The character "V" means finding eigenvalues and eigenvectors. "U" stands for referencing an upper triangular symmetric matrix and it will be overwritten with eigenvectors. n is the dimension of the matrix, w is the array to return the computed eigenvalues. The second n is the leading dimension of A.

6.4.3 Notes on Using BLAS and LAPACK

As we mix our use of FORTRAN and C++, seeming slight differences run the risk of unexpected pitfalls. Here are some points to be aware of when dealing with BLAS and LAPACK.

6.4.3.1 Column Major and Row Major Storage Formats for Matrices

A matrix is two-dimensional, but the memory of a computer is one-dimensional. Suppose we store the following matrix in memory,

$$A = \begin{pmatrix} 1 & 2 & 3 \\ 4 & 5 & 6 \end{pmatrix}.$$

The column major and row major formats differ in how they store matrices in memory. Figure 6.2 indicates that data are stored in column major format as follows:

$$1, 4, 2, 5, 3, 6.$$

Likewise, Fig. 6.3 indicates that data are stored in row major format as follows:

$$1, 2, 3, 4, 5, 6.$$

Column major is used in FORTRAN, MATLAB, and Octave. Row major is used in C and C++. This might become a problem when one is considering using continuous access of memory for fast (optimized) implementations.

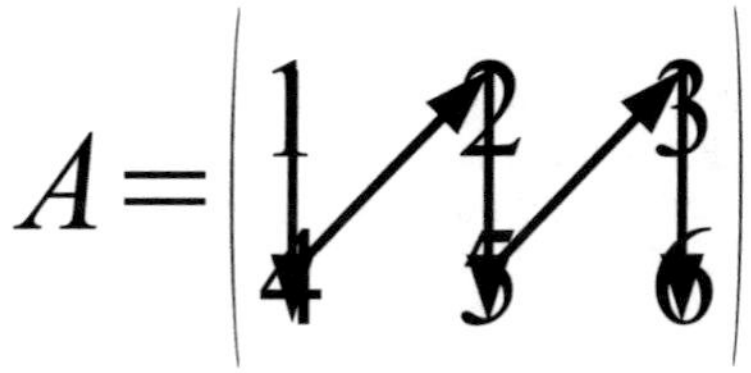

Fig. 6.2 Column major: How matrix data is stored in the memory in column major format

6.4.3.2 The Leading Dimension of a Matrix

An actual calculation will use the partition matrix or a sub-matrix of the original matrix. For this purpose, the "leading dimension" is required.

"LDA" and "LDB" are the leading dimensions of matrix A or B. In the example shown in Fig. 6.4, the $M \times N$ matrix A' is a partial matrix of the original matrix A whose size is $K(=LDA) \times L$. The pointer A' indicates the address of the P and Qth element of A.

$$A' = \&A(P + Q * K)$$

However, when we try to access the (i, j)-th element of A in C, the following access is apparently not correct.

$$A'(i + j * M)$$

It should be

$$A'(i + j * K).$$

In this case, "LDA" is equal to "K". Since A' is inside a larger matrix, to access the next column of the matrix, we need the leading dimension instead of the actual dimension so that we can access it like a usual matrix.

Let us show a more realistic example. The matrix–matrix product can be speedup on modern computers by calculating the product of the partition matrices instead of running a textbook algorithm [7].

Consider two $n \times n$ matrices, A and B and their product $C = AB$. A_{pq}, B_{qr}, and C_{pr} are partition matrices composing A, B and C.

Fig. 6.3 Row major: How matrix data is stored in the memory in row major format

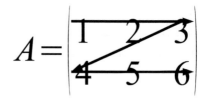

Fig. 6.4 Concept of leading dimension: an $M \times N$ matrix A' can be treated as a sub-matrix of a larger $K(=LDA) \times L$ matrix

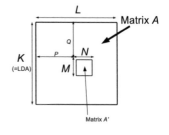

$$A = \begin{pmatrix} A_{11} & A_{12} & \cdots & A_{1q} \\ A_{21} & A_{22} & \cdots & A_{2q} \\ \vdots & \vdots & \ddots & \vdots \\ A_{p1} & A_{p1} & \cdots & A_{pq} \end{pmatrix}, \quad B = \begin{pmatrix} B_{11} & B_{12} & \cdots & B_{1r} \\ B_{21} & B_{22} & \cdots & B_{2r} \\ \vdots & \vdots & \ddots & \vdots \\ B_{q1} & B_{q1} & \cdots & B_{qr} \end{pmatrix}$$

$$C = \begin{pmatrix} C_{11} & C_{12} & \cdots & C_{1r} \\ C_{21} & C_{22} & \cdots & C_{2r} \\ \vdots & \vdots & \ddots & \vdots \\ C_{p1} & C_{p1} & \cdots & C_{pr} \end{pmatrix}$$

The following relations for the matrix–matrix product for pairs of partition matrices also hold[4]:

$$C_{ij} = \sum_k A_{ik} B_{kj}$$

To calculate C_{ij}, we use of concept of leading dimension. We access the elements of the partition matrices, such as A_{ij}, of the original A.

$$A_{ij}(p + q * lda)$$

By accessing like this, it becomes possible to handle the matrix–matrix product of partition matrices like an ordinary matrix product.

Arrays Start from 1 in FORTRAN and from 0 in C/C++

In FORTRAN, the array starts from 1, but in C and C++, it starts from 0.

- The loop variable is generally 1 to N (FORTRAN) or 0 to less than N (C, C++).
- Accessing the x_ith element of a vector is $X(I)$ in FORTRAN, but $x[i-1]$ in C.
- Accessing $A_{i,j}$th element of a matrix is $A(I, J)$ in FORTRAN, and $A[i - 1 + (j - 1) * lda]$ in C, in column major format.

6.5 Benchmark Results of Optimized BLAS and LAPACK

OpenBLAS is the definitive package for running BLAS and LAPACK optimized for recent CPUs. It is almost the fastest of all BLAS and LAPACK implementations available for computers including Linux distributions.[5] While we recommend using it as a black box, the reader may want to understand why it is fast. Here, we describe the idea of a bottleneck in computing and we show some examples in Ubuntu.

---

[4] It looks very similar to the textbook implementation. However, in this case, we use sub-matrices instead of numbers. This algorithm makes use of the hierarchical structure of the memory cache; it is also suitable for multicore CPUs because of the independence of each sub-matrix C_{pq}.

[5] The situation before 2010 was quite chaotic, because the source code was hidden by vendors.

Fig. 6.5 Conceptual
diagram of bottleneck. The
thinnest part, the neck, of the
bottle determines the water
flow

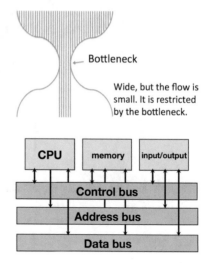

Fig. 6.6 Von Neumann type
computer conceptual
diagram

6.5.1 Bottlenecks in Computing

Suppose we had to perform some job consisting of three steps, A, B, and C. First, we would perform step A, then step B, then step C. Let us further assume that steps A and C could be improved very efficiently; what can we say about the efficiency of the work as a whole?

If steps A and C could be sped up so that it took almost no time, step B would become the bottleneck limiting the efficiency of the whole job, as illustrated in Fig. 6.5.

Programs run on computers. The idea is quite natural for us nowadays, but it was not in the earliest computers. Programmable (modern) computers are called von Neumann computer (Fig. 6.6). The speed of the CPU does not necessarily determine the speed of the computer. There is also the von Neumann bottleneck. If the "bus" connecting the CPU and memory or connecting the CPU and input/output, is not fast enough, it becomes a bottleneck. Usually, the speed of the bus is called the "bandwidth". In particular, the bandwidth of the CPU–memory bus is often called the memory bandwidth.

If you want to run a program faster, it is very important to find out where the bottleneck of the program is. Replacing the library with an optimized one usually does the job, but you should know more.

6.5.2 Theoretical Peak Performance of CPUs

Moore's law states that the integration level of transistors doubles every 18 months (Fig. 6.7).

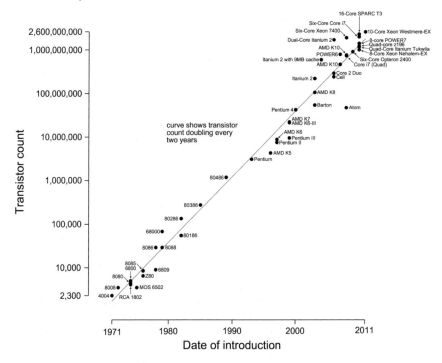

Fig. 6.7 Conceptual diagram of Moore's law (The figure is based on Wgsimon https://commons. wikimedia.org/wiki/File:Transistor_Count_and_Moore's_Law_-_2011.svg)

Since the speed of a computer is almost proportional to the number of transistors on the chip, Moore's law implies that the speed of CPUs also grows exponentially faster. Although many barriers have been encountered that would have limited this increase, Moore's law has been supported by physical and technological breakthroughs.

As you can see from Fig. 6.7, until about 2000, the performance of computers mainly increased by raising the operating frequency. However, the operating frequency has since reached its limit (about 4 GHz). Currently, it is commonplace to use a CPU packed with many cores, which is called a multicore CPU. As multicore technology advances, it is now not uncommon to have more than ten cores on a CPU. This means it is very important to support multicore technology when developing programs or libraries.

"Flops" means floating-point operations per second, or how many times a floating-point operation can be performed per second. Flops is used for measuring the performance of CPUs. For example, when performing one floating-point operation with one clock, the performance of a CPU operating at 1 GHz is 1 Gflops, calculated as follows:

$$1 \text{ flop} \times 1 \text{ clock} \times 1 \text{ G}/(\text{clocks}) = 1 \text{ Gflops}.$$

In the case of a four-core CPU operating at 1 GHz, for example, where each core of CPU performs one floating-point operation in 1 clock, the performance is 4 Gflops, calculated as follows:

$$1\,\text{flop} \times 1\,\text{clock} \times 1\,\text{G}/(\text{clocks}) \times 4\,\text{cores} = 4\,\text{Gflops}$$

Recent cores may contain single instruction, multiple data (SIMD) operations that perform a number of operations with one instruction. For example, Intel's SSE 2 instruction set allows four instructions (two double-precision additions and multiplications of floating-point arithmetic) to be executed at once.

Thus, the theoretical peak performance becomes 64 Gflops on an eight-core CPU[6] at 2 GHz with SSE2 enabled

$$4\,\text{flop} \times 1\,\text{clock} \times 2\,\text{G}/(\text{clocks}) \times 8\,\text{cores} = 64\,\text{Gflops}$$

To measure flops, we do not have to run something particularly meaningful; performing "1+1" on every arithmetic unit is enough to calculate the "theoretical peak performance". Here, we want to measure the peak performance of particular (meaningful) calculation such as matrix–matrix multiplication. In that case, the measured peak performance will not be as high as the theoretical peak performance. Because in real calculations, data transfers are required where in the CPU moves data from memory to the CPU and from the CPU to memory than just running the arithmetic units. We refer to such a measure as "the peak performance on matrix-matrix multiplications".

6.5.3 Memory Bandwidth and Byte Per Flop (B/F)

As you can see from the graph of memory bandwidth and CPU operation speed (Fig. 6.8), the CPU calculation speed has risen much faster than memory bandwidth. In fact, memory bandwidth has not increased much since about 1990.

Back then, optimizing a program to do something faster entailed putting the results or temporary results of a calculation in memory and using them as far as possible to avoid new calculations.

After 1990, the situation is completely different. Memory bandwidth has become narrower and narrower relative to floating point operations. As a result, algorithms now have to work without writing or reading memory as much as possible, and keeping results in registers or small cache memories.

Thus, although the memory bandwidth of the computer is a very important value, it is frequently overlooked. Despite that it is somewhat complicated, we have to look closely at three things: (1) the speed of memory transfers (DDR3/4/5), (2)

---

[6]The calculation becomes difficult when the clock changes dynamically such as in the case of TurboBoost.

Fig. 6.8 Graph of memory bandwidth and CPU operation speed (The figure is reproduced with permission from Dr. John McCalpin https://www.cs. virginia.edu/stream/)

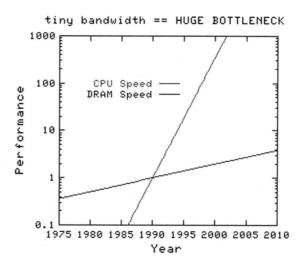

the configuration of the bus between the CPU and the memory module (how many memory channels equip the CPU), and (3) the bandwidth of the memory module and the motherboard. The theoretical peak performance of memory transfers is more difficult to calculate than the CPU performance. Thus, actually measuring it on your machine might be good enough.

We measured the memory bandwidth with Ubuntu 16.04 running on an Intel Xeon E5-2603 @ 1.80 GHz, DDR 3 800/1066 machine. Note that when you run this test, you may also have to change the -DSTREAM_ARRAY_SIZE value from 100,000,000, as it depends on the system (too small a value makes it difficult to measure, as noise becomes relatively larger).

```
$ wget https://www.cs.virginia.edu/stream/FTP/Code/stream.c
$ gcc -O3 -DSTREAM_ARRAY_SIZE=100000000 -fopenmp stream.c -o stream
$ ./stream
-------------------------------------------------------------
STREAM version $Revision: 5.10 $
-------------------------------------------------------------
This system uses 8 bytes per array element.
-------------------------------------------------------------
Array size = 100000000 (elements), Offset = 0 (elements)
Memory per array = 762.9 MiB (= 0.7 GiB).
Total memory required = 2288.8 MiB (= 2.2 GiB).
Each kernel will be executed 10 times.
 The *best* time for each kernel (excluding the first iteration)
 will be used to compute the reported bandwidth.
-------------------------------------------------------------
Number of Threads requested = 8
Number of Threads counted = 8
-------------------------------------------------------------
Your clock granularity/precision appears to be 1 microseconds.
Each test below will take on the order of 31033 microseconds.
   (= 31033 clock ticks)
Increase the size of the arrays if this shows that
you are not getting at least 20 clock ticks per test.
-------------------------------------------------------------
WARNING -- The above is only a rough guideline.
For best results, please be sure you know the
precision of your system timer.
```

```
----------------------------------------------------------------
Function     Best Rate MB/s  Avg time      Min time      Max time
Copy:             34833.3    0.046317      0.045933      0.047808
Scale:            35946.3    0.045001      0.044511      0.045305
Add:              38871.4    0.062394      0.061742      0.064245
Triad:            38862.1    0.062550      0.061757      0.063273
----------------------------------------------------------------
Solution Validates: avg error less than 1.000000e-13 on all three arrays
----------------------------------------------------------------
```

The first operation is a simple copy a(i)=b(i), the second scalar multiplication a(i)=q*b(i). Add means vector addition a(i)=b(i)+c(i), and Triad means addition of two vectors, with one multiplied by a constant a(i)=b(i)+q*c(i) (this is the same as DAXPY Level 1 BLAS routine). As you can see, the memory bandwidth of these operations runs from about 35 GB/s–40 GB/s.

In this case, it is also important to use multithreading on a multicore computer. Although memory bandwidth looks unrelated to multicore computations, it becomes important when the bandwidth is almost used up, because if one core fully occupies the memory bandwidth, the other cores will not be able to use the memory.

Finally, we explain the terminology "byte per flop". There are two meanings of this phrase. One is for algorithms (or programs), the other for computers. The B/F of an algorithm is defined as the necessary memory access amount in bytes.

For example, let us calculate the B/F of the triad operation: $y(i) + a \times x(i)$, where x and y are n-dimensional vectors and a is a scalar.

$$y_i \leftarrow y_i + ax_i.$$

Here, we need $2n$ floating-point arithmetic operations, $2n$ data reads, and n writes. One double-precision number is 8 bytes; therefore, the B/F value is

$$8(2n + n)/2n = 3 * 8/2 = 12$$

The B/F of a computer is defined as how many memory accesses are made, in units of bytes, in one floating-point operation.

For example, an Intel Xeon E5-2603 @ 1.80 GHz (four cores) with a theoretical peak performance of 101.25 Gflops and memory bandwidth of approximately 40 GB/s has a B/F of

$$40/101.25 = 0.4.$$

B/F decreases when the memory bandwidth is smaller than the computation speed. Faster processing is thus achieved by widening the memory bandwidth.

A machine with a small B/F cannot run large B/F programs efficiently; its memory bandwidth would be almost completely occupied, even though its CPU may be hardly working. This problem is an ongoing one, so readers should think about the B/Fs of their programs. That is, you can waste your CPU operations, but you should never waste memory bandwidth.

6.5.4   Example of Installation and Selection of Libraries in Ubuntu

Let us investigate the BLAS and LAPACK packages included by default in Ubuntu 16.04.

```
$ apt-cache search libblas
libblas-common - Dependency package for all BLAS implementations
libblas-dev - Basic Linear Algebra Subroutines 3, static library
libblas-doc - Transitional package for BLAS manpages
libblas3 - Basic Linear Algebra Reference implementations, shared library
libatlas-base-dev - Automatically Tuned Linear Algebra Software, generic
    static
libatlas3-base - Automatically Tuned Linear Algebra Software, generic
    shared
libblas-test - Basic Linear Algebra Subroutines 3, testing programs
libblasr - tools for aligning PacBio reads to target sequences
libblasr-dev - tools for aligning PacBio reads to target sequences (
    development files)
libopenblas-base - Optimized BLAS (linear algebra) library (shared library)
libopenblas-dev - Optimized BLAS (linear algebra) library (development
    files)
$ apt-cache search liblapack
liblapack-dev - Library of linear algebra routines 3 - static version
liblapack-doc - Library of linear algebra routines 3 - documentation (HTML)
liblapack-doc-man - Library of linear algebra routines 3 - documentation (
    manual pages)
liblapack3 - Library of linear algebra routines 3 - shared version
liblapacke - Library of linear algebra routines 3 - C lib shared version
liblapacke-dev - Library of linear algebra routines 3 - Headers
libatlas-base-dev - Automatically Tuned Linear Algebra Software, generic
    static
libatlas3-base - Automatically Tuned Linear Algebra Software, generic
    shared
liblapack-pic - Library of linear algebra routines 3 - static PIC version
liblapack-test - Library of linear algebra routines 3 - testing programs
libopenblas-base - Optimized BLAS (linear algebra) library (shared library)
libopenblas-dev - Optimized BLAS (linear algebra) library (development
    files)
```

The reference implementations of BLAS and LAPACK should give the correct results (otherwise, there are bugs of the libraries and/or compilers). Automatically Tuned Linear Algebra Software (ATLAS) [8], developed by R. Clint Whaley and coworkers, is intended to reduce the cost of hand tuning on individual CPUs and architectures; this package includes sets of parameters and implementations for the most important parts (called kernels), and guidelines on what to choose when building a library. The strategy for choosing the fastest one is called automatic tuning.

OpenBLAS is an unofficial successor or a fork of GotoBLAS2 [9]. Goto Kazushige had been developing hand-tuned high-speed BLAS and LAPACK routines for several years, but stopped in 2010. The final version, GotoBLAS2 1.13, was released with an open-source license, called BSD. The development has since been taken over by Zhang Xianyi [10]. OpenBLAS now supports recent Intel, AMD, POWER, and ARM CPUs. Even though hand tuning is mainly used, some techniques of automatic tuning are also used.

Since the auto-tuning strategy is used in ATLAS, it is necessary for users to build ATLAS. Moreover, the package provided by Ubuntu is slow; it does not use multicore technology or SSE4. Comparing it with other packages would be unfair.

The libraries are installed as follows:

```
$ sudo apt-get install libblas-dev libatlas-base-dev libopenblas-dev \
liblapack-dev libatlas-base-dev libopenblas-dev
```

The reference (2), atlas (1), and openblas (0, 3) versions of BLAS and LAPACK can be chosen.

6.5.5 Performance of Different Implementations of GEMM

Let us compare the performances of the DGEMM, which is the matrix–matrix product, implementations of the different libraries. Let us consider the case of $C \leftarrow \alpha AB + \beta C$, where A, B, and C are $n \times n$ matrices, and α and β are scalars.

$$(C)_{ij} \leftarrow \alpha \sum_{k}^{n} (A)_{ik}(B)_{kj} + \beta(C)_{ij}$$

In this case, we can make full use of the CPU, because the B/F of matrix–matrix multiplication is almost 0 for large n. Let us prove this. The number of elements to be read is $3n^2$ (for A, B and C), and the number of elements to be written is n^2 (for only C). To calculate the (i, j)th element of C, C_{ij}, we need $n + 2$ multiplications and $n - 1$ additions (Fig. 6.9). Since this calculation should be done n^2 times, $2n^3 + n^2$ operations are required to calculate C. When the value of n is large, the amount of memory used is about $4n^2$, and the operations on it number $2n^3$. Therefore, B/F is

$$\frac{8 \times 4n^2}{2n^3} \rightarrow \frac{16}{n}$$

Apparently therefore, B/F approaches 0 when n is large. The B/F of the CPU is strictly larger than 0; therefore, GEMM runs very fast on most computers.

Next, we show the results of Octave on Ubuntu 16.04; the CPU was a four-core Intel Xeon E5-2603 @ 1.80 GHz. First, we show the result of OpenBLAS;

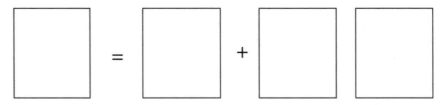

Fig. 6.9 Conceptual figure of DGEMM

```
$ octave
GNU Octave, version 4.0.0
Copyright (C) 2015 John W. Eaton and others.
This is free software; see the source code for copying conditions.
There is ABSOLUTELY NO WARRANTY; not even for MERCHANTABILITY or
FITNESS FOR A PARTICULAR PURPOSE.  For details, type ''warranty".

Octave was configured for ''x86_64-pc-linux-gnu''.

Additional information about Octave is available at http://www.octave.org.

Please contribute if you find this software useful.
For more information, visit http://www.octave.org/get-involved.html

Read http://www.octave.org/bugs.html to learn how to submit bug reports.
For information about changes from previous versions, type 'news'.

octave:1> n=4000; A=rand(n); B=rand(n);
octave:2> tic(); C=A*B; t=toc(); GFLOPS=2*n^3/t*1e-9
GFLOPS =    101.25
octave:3>  ,
```

then of ATLAS,

```
octave:1> n=4000; A=rand(n); B=rand(n);
octave:2> tic(); C=A*B; t=toc(); GFLOPS=2*n^3/t*1e-9
GFLOPS =   6.4421
```

finally, of reference BLAS,

```
octave:1> n=4000; A=rand(n); B=rand(n);
octave:2> tic(); C=A*B; t=toc(); GFLOPS=2*n^3/t*1e-9
GFLOPS =   1.2806
```

OpenBLAS was the fastest at 101.25 Gflops. The theoretical peak performance of the Xeon E5-2603 @ 1.80 GHz is 115.2 Gflops, calculated as follows[7]

$$1.8\,\text{GHz} \times 4\,\text{cores} \times 8\,(\text{AVX}) = 115.2\,\text{Gflops}.$$

The actual performance is thus about 88% of the theoretical peak.

6.5.6 Comparison of GEMV Performances

Next, let us compare the different implementations of DGEMV, the matrix–vector product. Let us consider the following case (Fig. 6.10).

$$y_i \leftarrow \alpha \sum_{j}^{n} (A)_{ij} x_j + \beta y_j,$$

---

[7] AVX refers to Intel Advanced Vector Extensions which is an extension of the SIMD-type instructions succeeding SSE. It has a 256 bit width and can calculate additions and multiplications in one clock. It can store four double-precision values in 256 bits. It can calculate two multiplications per clock, so it is possible to perform eight operations in 1 clock.

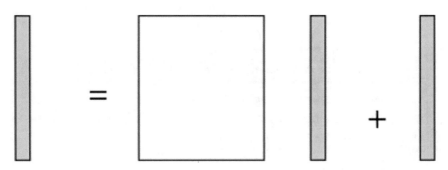

Fig. 6.10 Conceptual diagram of DGEMV

where A is an $n \times n$ matrix and x and y are n-dimensional vectors. Let us compute the B/F of GEMV. The matrix and vector data are read and written $n^2 + 2n$ and n times, respectively. There are $n + 1$ multiplications and $n - 1$ additions for each element of αAx, and 1 multiplication for βy. Finally, there is 1 more addition to calculate $\alpha Ax + \beta y$. Thus, there are totally $n^2 + 3n$ operations for reading and writing to memory and $2n^2 + 2n$ operations involved in calculating the new vector y. Therefore, for large n, the B/F of DGEMV is

$$\frac{8(n^2 + 3n)}{2n^2 + 2n} \to 4.$$

In other words, the order of memory usage is the same as that of the calculations. Data is discarded after read and processed. And, it will never be reused after doing some operations. In this case, DGEMV is not fast if only the CPU is powerful.[8]

Next, we show the results of Octave on Ubuntu 16.04; the CPU used was an Intel Xeon CPU E5-2603 @ 1.80 GHz. Since the maximum measured memory bandwidth is about 40 GB by the STREAM bench (see Sect. 6.5.3), 5 Gflops would be is the maximum performance (40 GB/8B = 5G), which is 20 times slower than DGEMM's 100 Gflops. This means the CPU was hardly used at all during the calculation.

```
$ octave
octave:1> n=10000; A=rand(n); y = rand(n,1) ; x = rand(n,1) ; tic();
       y=A*x; t=toc(); GFLOPS=2*n^2/t*1e-9
GFLOPS =   5.4217
```

It is clear that the peak memory bandwidth was obtained.
In ATLAS, we obtained 0.99 Gflops as follows.

```
$ octave
octave:1> n=10000; A=rand(n); y = rand(n,1) ; x = rand(n,1) ;
tic(); y=A*x; t=toc(); GFLOPS=2*n^2/t*1e-9 GFLOPS =   0.98827
```

---

[8]However, it is better to use the Xeon because it has more memory bandwidth despite it being more expensive to run on a Core i7.

Its performance is poorer than that of Octave because it does not take full advantage of memory bandwidth. It supports only one core (so that it will run on all computers). Thus, this comparison is unfair.

Reference BLAS, we obtained 1.26 Gflops as follows.

```
octave:1> n=10000; A=rand(n); y = rand(n,1) ; x = rand(n,1) ;
tic(); y=A*x; t=toc(); GFLOPS=2*n^2/t*1e-9 GFLOPS =   1.2627
```

This result is still very slow, but slightly faster than ATLAS. Although we didn't analyze the result, we think it is due to the use of a different optimizer from that of ATLAS. The ATLAS binary was optimized for general environment, which must support older (thus slower) architectures.

6.6   Summary

Starting with the history of linear algebra, we showed how important linear algebra has become and how it is used everywhere in today's technology. We stressed the usefulness of pre-developed libraries because they are much easier to use. They deal well with numerical errors and have machine-efficient algorithms.

After that, we described the BLAS and LAPACK linear algebra libraries. We showed some examples of using these libraries by calling from C++ and how to use optimized BLAS and LAPACK for Ubuntu, a distribution of Linux. We then gave an overview of the optimized BLAS and LAPACK together with examples of using Octave. We briefly explained the concepts of bottleneck, flops, and byte per flop of machines and applications and how to measure the theoretical peak performance.

Finally, we showed examples of linear algebra calculations. Two extreme cases were examined, DGEMM and DGEMV. We should exploit CPU performance on DGEMM and memory bandwidth on DGEMV.

The readers should take care of bottlenecks and the byte per flop of their programs. We would like to conclude this chapter with the hope that it has helped readers to make their programs faster.

Exercises

1. Find the theoretical performance of double-precision operations of a machine equipped with two Intel Xeon X5680s. Suppose the operating frequency is 3.3 GHz and Turbo Boost is not used.
2. Figure 6.11 is a performance graph of the matrix–matrix product of the machine listed above for the DGEMM comparison. The horizontal axis is the size of the matrix, and the vertical axis is the performance in Mflops. Read the maximum performance from the graph for when the size of the matrix is infinity and find the percentage of maximum performance of the CPU that is attained.
3. Figure 6.12 is the same calculation as before, and it is an enlargement of part of the figure. Why is the curve "jagged"?

Fig. 6.11 Illustration of the performance of Intel Xeon X5680 in matrix–matrix multiplication of square matrices of various dimensions using Intel MKL (math kernel library) The horizontal axis shows dimension, and the vertical axis shows Mflops

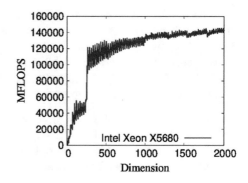

Fig. 6.12 An enlargement of a part of the Fig. 6.11

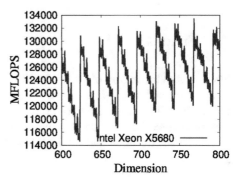

References

1. Author unknown, The nine chapters on the mathematical art, around the 1st century BC to the second century AD
2. IEEE, IEEE standard for floating-point arithmetic, IEEE Std 754-2008, pp. 1–70 (2008)
3. N.J. Higham, *SIAM: Society for Industrial and Applied Mathematics*, 2nd edn. (2002)
4. S. Hauberg, J. W. Eaton, D. Bateman, *GNU Octave Version 3.0.1 Manual: A High-level Inter-active Language for Numerical Computations* (CreateSpace Independent Publishing Platform, 2008)
5. MATLAB, version 7.10.0 (R2010a). The MathWorks Inc., Natick, Massachusetts (2010)
6. BLAS quick reference card, http://www.netlib.org/blas/blasqr.pdf
7. B. Kågström, P. Ling, C. Van Loan, GEMM-based level 3 BLAS: high-performance model implementations and performance evaluation benchmark, ACM Trans. Math. Softw. **24**(3), 268–302 (1998)
8. R.C. Whaley, J.J. Dongarra, in *Proceedings of the 1998 ACM/IEEE Conference on Supercomputing, SC '98, 1* (IEEE Computer Society, Washington, 1998)
9. K. Goto, R.A. van de Geijn, ACM Trans. Math. Softw. **34**, 12:1 (2008)
10. X. Zhang, Q. Wang, Y. Zhang, in *IEEE 18th International Conference on Parallel and Distributed Systems (ICPADS)*, vol. 17 (IEEE Computer Society, 2012)

Chapter 7
High-Performance Algorithms for Numerical Linear Algebra

Yusaku Yamamoto

Abstract Matrix computations lie at the heart of many scientific computations. While sophisticated algorithms have been established for various numerical linear algebra problems such as the solution of linear simultaneous equations and eigenvalue problems, they require considerable modification with the advent of exaFLOPS- scale supercomputers, which are expected to have a huge number of computing cores, deep memory hierarchy, and increased probability of hardware errors. In this chapter, we discuss projected hardware characteristics of exaFLOPS machines and summarize the challenges to be faced by numerical linear algebra algorithms in the near future. Based on these preparations, we present a brief survey of recent research efforts in the field of numerical linear algebra targeted at meeting these challenges.

7.1 Introduction

In the past few decades, the performance of supercomputers has been growing at a rate of 1000 times per decade. If this rate continues, exaFLOPS machines, which are capable of performing 10^{18} floating-point operations per second, will be available around 2020. It is expected that exaFLOPS machines will have nearly 10^9 processing cores and very deep memory hierarchy, which exacerbates the discrepancy between floating-point performance and memory throughput. In addition, the failure rate can increase due to the increasing number of components. To cope with such problems and exploit the potential performance of exaFLOPS machines, we need drastic changes on the side of numerical algorithms. In this chapter, we focus on numerical linear algebra algorithms and discuss the challenges in the exaFLOPS era. We also highlight some of the recent research efforts targeted at meeting these challenges.

This chapter is structured as follows: Sect. 7.2 outlines projected hardware characteristics of exaFLOPS machines based on the "HPCI Technical Roadmap White Paper", which was published in Japan in 2012. In Sect. 7.3, the requirements on future numerical linear algebra algorithms are discussed from the application side.

Y. Yamamoto (✉)
The University of Electro-Communications, Chofu, Japan
e-mail: yusaku.yamamoto@uec.ac.jp

© Springer Nature Singapore Pte Ltd. 2019
M. Geshi (ed.), *The Art of High Performance Computing for Computational Science, Vol. 1*, https://doi.org/10.1007/978-981-13-6194-4_7

Based on these preparations, we point out several challenges faced by numerical linear algebra algorithms in the exaFLOPS era in Sect. 7.4. Some of the research efforts being conducted to meet these challenges are highlighted in Sect. 7.5. Finally, Sect. 7.6 gives some conclusion.

7.2 Hardware Characteristics of ExaFLOPS Machines

7.2.1 Supercomputers in Year 2020

The "HPCI Technical Roadmap White Book" [26] classifies supercomputers in 2018 into four categories, namely, (1) general purpose type, (2) memory capacity- and bandwidth-oriented type, (3) FLOPS-oriented type, and (4) reduced memory type. Here, the general purpose type is an enhanced version of today's supercomputers, for which all of the memory capacity, data transfer throughput and floating-point performance is equally enhanced. This type of supercomputers can be applied to solve a wide variety of problems, like the K computer [29] in Japan, which is one of today's representative general purpose supercomputers.

In the case of memory capacity- and bandwidth-oriented machines, more hardware resources are allocated for increasing the memory bandwidth, at the cost of reducing the floating-point performance. In contrast, in the case of FLOPS-oriented machines, more resources are allocated for increasing the floating-point performance, by sacrificing the memory bandwidth. The reduced memory machines have only on-chip memory, which is extremely fast but has limited capacity. The White book compares the floating-point performance achievable by each type of supercomputers, assuming power consumption and footprint that are equal to those of the K computer (20 MW and 2,000–3,000 m^2, respectively). The result is 200–300 PFLOPS for the general-purpose type, 50–100 PFLOPS for the memory capacity- and bandwidth-oriented type, 1,000–2,000 PFLOPS for the FLOPS-oriented type, and 500–1,000 PFLOPS for the reduced memory type ([26, Table 2-5]). Hence, the leading candidates of exaFLOPS machines are the FLOPS-oriented and reduced memory types. However, the total memory capacity of the reduced memory type is estimated as 0.1–0.2 PBytes, which is far less than that of the FLOPS-oriented type, estimated as 5–10 PBytes. This severely limits the kind of applications that can be executed on reduced memory machines. Based on this consideration, we mainly focus on the FLOPS-oriented machines in this chapter.

7.2.2 Hardware Characteristics of FLOPS-oriented Supercomputers

In the following, we outline the projected hardware specifications of exaFLOPS machines, assuming the FLOPS-oriented type, and point out their important features. A typical example of an exaFLOPS machine assumed in this chapter is shown in Fig. 7.1.

Parallelism of 10^9 order

At present, the clock frequency of CPU cores is at most a few GHz. This situation will not change drastically in the near future, mainly due to the requirement of keeping the power consumption at an affordable level. This means that we need 10^9 order of parallelism to achieve exaFLOPS. This parallelism will be realized by a hierarchy, consisting of instruction level, core level, chip level, and node level parallelism.

Deep memory hierarchy

Today's supercomputers already have a fairly deep memory hierarchy, consisting of on-chip registers, several levels of on-chip and off-chip cache, main memory within a node, and main memory in other nodes. This hierarchy will become even deeper and more complicated in exaFLOPS machines, corresponding to the hierarchical parallelism stated above.

Increase in the data transfer cost

Up to now, the floating-point performance of supercomputers has been increasing more rapidly than memory access performance or internode communication performance. This has resulted in a severe discrepancy between the computation speed and the data transfer speed, which is expected to grow even larger in the future. Let us divide the data transfer performance into latency and throughput and consider them separately. According to the prediction in [26], a FLOPS-oriented machine

Fig. 7.1 Architecture of an exaFLOPS machine presupposed in this chapter

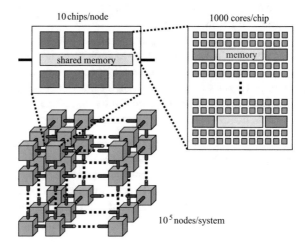

10 chips/node

1000 cores/chip

shared memory

memory

10^5 nodes/system

Fig. 7.2 AllReduce
operation using a binary tree

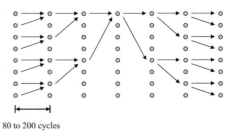

80 to 200 cycles

with total performance of 1,000–2,000 PFLOS will have a total memory bandwidth of 5–10 PBytes/s. Hence, the ratio of data transfer throughput to the floating-point performance is 0.005 Byte/FLOP. This means that we need to perform at least 1600 operations on each double-precision data (8 bytes) fetched from memory in order to fully exploit the machine's floating-point performance. In contrast, for the K computer, the total performance is 10 PFLOPS and the total memory bandwidth is 5 PByte/s, so the ratio is 0.5 Byte/FLOP. Thus, the relative memory access cost of a FLOPS-oriented machine is 100 times higher than that of the K computer. As for the latency, [26, Table 2-3] estimates the inter-core synchronization/communication latency as 100 ns ($\sim$100 cycle) and the internode communication latency as 80–200 ns. This means that virtually no performance enhancement can be expected with respect to the latency. Considering that exaFLOPS machines have much higher floating-point performance and larger number of nodes than today's supercomputers, we can conclude that the effect of latency on execution time will be far more serious on exaFLOPS machines. The effect of latency will be the most salient in AllReduce-type communication such as arises in the inner product of two vectors. For example, an AllReduce operation among 10^5 nodes using a binary tree will require several thousand cycles (Fig. 7.2).

The problem of power consumption

For exaFLOPS machines, reducing the power consumption is an important issue not only from environmental considerations, but also to suppress heat generation. In considering power consumption, we must note that off-chip data transfer consumes large power. In fact, according to the prediction in [26], the power consumption for one double-precision arithmetic operation is only 1.1 pJ/FLOPS [26, Table 2-1], while it is 2 pJ for on-chip data transfer and as much as 200 pJ for off-chip data transfer [26, Sect. 2.1.4]. It is therefore critical to reduce off-chip data transfer as much as possible not only for higher performance but also for lower power consumption.

Increase in the failure rate due to the increasing number of components

In the exaFLOPS era, the process rule of LSIs will be much finer, and at the same time, the number of components that constitute a machine will be much larger than today. As a result, the possibility of a failure, either due to soft errors caused by alpha particles or due to the breakdown of a component, is expected to increase.

7.3   Changes of the Requirements from the Application Program Side

So far, we have discussed the characteristics of exaFLOPS machines, which can have an impact on the design of future numerical linear algebra algorithms. On the other hand, the requirements on these algorithms from the application side can also change in the exaFLOPS era.

From weak scaling to strong scaling
Up to this day, scientific computing has evolved in the direction of solving larger and larger problems using more and more computing power and memory capacity. Accordingly, in the world of parallel processing for scientific computing, much emphasis has been laid on *weak scaling*, where the problem size is increased in proportion to the number of computing cores used. However, the sizes of problems that can be solved with today's supercomputers are already fairly large, and not all applications require expanding the problem size further. For example, for molecular dynamics applications, there is a strong need for large scale computation in the time direction, in which the system size is fixed and the number of time steps is extended to, say, 1 million steps [34]. In such a case, because parallelizing in the time direction is difficult,[1] one has to shorten the computation of one time step as much as possible in order to finish the whole computation within practical time. Accordingly, it is required to use the power of parallel processing not for solving a larger problem, but for solving a given problem in a shorter time. Thus, parallel performance under *strong scaling*, in which the problem size is fixed and the number of core is increased, becomes important.

Requirement of reducing the order of computational work
For many numerical linear algebra algorithms, the computational work grows faster than linearly with the matrix dimension n. In particular, for dense matrix computations such as the LU, QR, and singular value decompositions, the computational work typically grows as $O(n^3)$, or as $O(d^{1.5})$, where $d = O(n^2)$ is the data size of the input matrix. In this case, even if we increase the number of cores in proportion to the data size d, the computational time grows as $O(n)$ or $O(d^{0.5})$, and for extremely large problems, we may not be able to finish the computation in affordable time. Hence, there is a need for algorithms with smaller order of computational work, possibly at the cost of introducing approximation in some sense.

Need for higher precision arithmetic
This is actually not a requirement from the application side, but from the numerical linear algebra algorithms themselves. If the matrix dimension is extremely large, say, $n \geq 10^6$, conventional double-precision arithmetic may not give sufficiently accurate results. In such a case, it is necessary to use higher precision arithmetic or reconsider the algorithm to attain the desired accuracy.

---

[1] Note, however, that there are efforts to parallelize in the time direction, like the famous parareal algorithm [1].

7.4   Challenges to Numerical Linear Algebra Algorithms

Based on the discussion in the preceding two sections, we can summarize the challenges faced by numerical linear algebra algorithms in the exaFLOPS era as follows.

(1) **10^9 order parallelism and adaptability to the hierarchical hardware**
    To fully exploit the computing power of exaFLOPS machines, algorithms must have 10^9 order parallelism. In addition, it is desirable that the algorithms have hierarchical parallelism, in conformity with the hierarchical parallelism of the hardware.

(2) **Reducing the amount of data transfer**
    On exaFLOPS machines, the cost of data transfer such as main memory access and internode communication is much larger than that of computation. It is therefore necessary to minimize the amount of data transfer. In the memory hierarchy, upper level storages are faster but have limited capacity, while lower level storages have larger capacity but lower data transfer speed. Thus, by increasing data reusability and performing as much computation on data as possible while it is in an upper level storage like cache, we can minimize data transfer from the lower level storage and improve the performance (Fig. 7.3). This will also help reduce power consumption.

(3) **Reducing the number of data transfer**
    It is also important to reduce the fixed cost incurred for each data transfer regardless of the data size. This includes the startup time of internode communication and the additional latency that occurs when accessing discontinuous memory locations. To keep this kind of costs to a minimum, we need to make the granularity of the algorithm as large as possible, by aggregating synchronizations and data transfers. It is also important to make memory access contiguous by choosing judicious data layout.

(4) **Fault tolerance**
    Basically, it is the role of the operating system to cope with hardware errors such as memory and communication errors. In addition, general-purpose methods such as checkpointing will be useful for recovering from hardware errors.

Fig. 7.3 Reducing the amount of data transfer by increasing data reusability

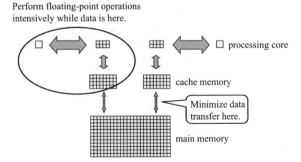

However, if the algorithm itself has some kind of redundancy and can continue the computation even in the presence of computation or communication errors, it will be very advantageous for the user.

(5) **Algorithms with reduced order of computational work**
One possible approach to solve extremely large problems within affordable computational time is to introduce some approximation into an algorithm and reduce the order of computational work at the cost of admitting a prescribed level of (possibly stochastic) errors. Algorithms based on low-rank approximation of tensors and stochastic algorithms fall into this category. In the case of big data analysis, the input matrix is usually obtained by sampling from some population, and therefore already contains sampling errors. So, when computing its singular value decomposition, for example, it seems unnecessary to seek for a solution that is accurate to the roundoff error level. In such applications, stochastic algorithms will be of choice.

(6) **Algorithms that are efficient under strong scaling conditions**
While much emphasis has been laid on weak scaling in the field of parallel processing for scientific computing, algorithms that are efficient in the sense of strong scaling will become more and more important in the exaFLOPS era. Under strong scaling conditions, communication/synchronization times and memory access latency will be dominant. Accordingly, development of algorithms that can reduce these overheads will become an important issue. Large grain algorithms are advantageous also in this regard.

(7) **Algorithms that can use higher precision arithmetic efficiently**
In large scale scientific computation in the exaFLOPS era, situations where double-precision arithmetic is inadequate will occur more often than now. However, performing all arithmetic operations in, say, quadruple precision will be too costly. Thus, algorithms using mixed precision arithmetic, which perform only critical parts using higher precision arithmetic, will be desirable. In addition, techniques for minimizing the influence of rounding errors by, for example, changing the order of arithmetic operations dynamically depending on the input data, will become important.

7.5 Research Trends of Numerical Linear Algebra Algorithms

In this section, we highlight some recent research efforts in the field of numerical linear algebra that are targeted at meeting the challenges stated in the previous section.

7.5.1 Sufficient Parallelism for Using 10^9 Cores Efficiently

It is mandatory for algorithms in the exaFLOPS era to possess enough parallelism to use 10^9 processing cores efficiently. We discuss this point for dense matrix algorithms and sparse matrix algorithms separately.

7.5.1.1 Dense Matrix Algorithms

Algorithms for square matrices

Let us consider typical dense matrix algorithms such as the LU and QR decompositions, tri-diagonalization for the symmetric eigenvalue problem, and Hessenberg reduction for the nonsymmetric eigenvalue problem [20]. These algorithms have a sequential nature, where the columns (or rows) are annihilated one by one at each step. The total computational work is $O(n^3)$, while that of each step is $O(n^2)$. Thus, when $n = 10^6$, even if there are 10^9 cores, each core has 10^3 operations to perform. However, in these algorithms, the pivot rows and columns for elimination have to be generated and broadcast at each step and this can become a performance bottleneck. To avoid this, scheduling techniques, which overlap these computations and communications with the elimination operation and hide the execution times of the former, become important. Recently, scheduling based on DAG (Directed Acyclic Graph) has been studied actively and emerges as one of the standard techniques [5]. In the DAG scheduling, tasks such as the generation of pivots and elimination are expressed as nodes and their dependency is expressed by a DAG. Then, a general purpose scheduling algorithm is applied to execute the tasks in the shortest time. Software named DAGuE has been developed and its effectiveness has been confirmed for various dense matrix algorithms like the LU and QR decompositions [9].

Algorithms for rectangular matrices and band matrices

In scientific applications, algorithms for rectangular matrices and band matrices also play an important role. Widely used algorithms of this category include the LU decomposition of a band matrix, tri-diagonalization of a band matrix, QR decomposition of a tall and skinny matrix and incremental orthogonalization, where the vectors to be orthogonalized is given one by one. In the following, we consider an $n \times b$ rectangular matrix or an $n \times n$ band matrix with half bandwidth b (Fig. 7.4). Then, typical computational work of these algorithms is $O(nb^2)$ or $O(n^2b)$, depending on the algorithm. The parallelism at each step is $O(b^2)$ for band LU decomposition

Fig. 7.4 Band matrix (left) and tall-skinny matrix (right)

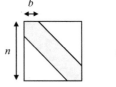

and band tri-diagonalization, $O(nb)$ for tall-skinny QR decomposition, and $O(n)$ for incremental orthogonalization (assuming that the modified Gram–Schmidt algorithm is used). These are too small for massively parallel environments.

Consequently, algorithms with larger parallelism have been actively studied for these problems. For example, there are several divide-and-conquer- type algorithms for band LU decomposition, which partition the matrix into several submatrices, compute the decomposition of each submatrix independently, and combine them to get the final results [24, 36, 44]. Among them, the so-called SPIKE algorithm is widely used and is implemented also in cuSPARSE [7], which is a sparse matrix library for NVIDIA GPUs. The textbook [19] discusses parallel algorithms for band LU decomposition in great detail.

For the QR decomposition of a tall and skinny matrix, the TSQR (Tall-Skinny QR) algorithm [8] based on the divide-and-conquer strategy has been proposed. In this algorithm, the input matrix A is divided horizontally into two submatrices A_1 and A_2 (Fig. 7.5), and their QR decompositions are computed independently. Next, the two upper triangular factors $R_1^{(1)}$ and $R_2^{(1)}$ are concatenated, and the QR factorization of the resulting matrix $R^{(1)}$ is computed. Then, the upper triangular factor of $R^{(1)}$ is the upper triangular factor of A. The orthogonal factor of A can be calculated as a product of the orthogonal factors at the first and the second stages. In this algorithm, most of the computational work is performed at the first stage, which can be computed completely independently for A_1 and A_2. Hence, the algorithm has a large grain parallelism. Furthermore, the TSQR algorithm can be applied to the QR decompositions of A_1 and A_2 recursively, enabling the use of 2^p processors, $p = 1, 2, \ldots$. The TSQR algorithm has almost the same numerical stability as the conventional Householder QR algorithm [32] and performance comparison of these two algorithms is conducted in various computational environments [17].

There is another algorithm for the tall-skinny QR decomposition namely, the CholeskyQR2 algorithm based on the Cholesky decomposition of the Gram matrix $A^\top A$ [16, 46]. It is even more suited to high-performance computing than the TSQR algorithm, since all the operation can be done in the form of level-3 BLAS. However, it can be used only when the condition number of A is less than 10^8 (when using double-precision arithmetic). To mitigate this limitation, an improved algorithm using shifted Cholesky decomposition has also been proposed [47].

Fig. 7.5 QR decomposition of a tall and skinny matrix by the TSQR algorithm

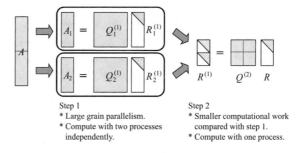

Step 1
* Large grain parallelism.
* Compute with two processes independently.

Step 2
* Smaller computational work compared with step 1.
* Compute with one process.

As algorithms for incremental orthogonalization, classical Gram–Schmidt (CGS) and modified Gram–Schmidt (MGS) methods have been widely used. As an alternative, there is an algorithm based on the compact WY representation of Householder transformations [43, 45], which requires twice the work but is unconditionally stable and has larger parallelism ($O(nb)$) than MGS. The TSQR algorithm can also be used as an incremental orthogonalization algorithm.

7.5.1.2 Sparse Matrix Algorithms

In sparse matrix computations, algorithms based on the projection method are widely used for the solution of linear simultaneous equations, eigenvalue computation, and computation of matrix functions $f(A)\mathbf{b}$. The examples include Krylov subspace methods for linear equations and Lanczos, Arnoldi and Jacobi–Davidson methods for eigenvalue problems. In these algorithms, the subspace used for projection is expanded by one dimension at each step. Since this operation is inherently sequential, parallelization is limited to within each step. Usually, the computational work of each step is dominated by matrix–vector multiplication, so the available parallelism is only $O(nz)$, where nz is the number of nonzero elements of the input matrix.

To increase the parallelism, several approaches have been proposed so far. One of them is block Krylov methods, which use multiple right-hand side vectors or multiple starting vectors. We will discuss these methods in the next subsection. There are other methods, which construct a numerical filter that passes only those vectors belonging to a specified subspace and use it to extract the desired subspace from the input vectors all at once. Filter diagonalization methods [42] and Sakurai–Sugiura methods [37] for eigenvalue computation fall into this category. In the following, we explain the idea of the Sakurai–Sugiura method, using the generalized eigenvalue problem $A\mathbf{x} = \lambda B\mathbf{x}$ as an example.[2] Suppose that we want to compute the eigenvalues $\lambda_1, \lambda_2, \ldots, \lambda_m$ within a closed Jordan curve Γ_1 on the complex plane (Fig. 7.6). By letting L and M be two natural numbers such that $LM \geq m$ and γ_1 be a point in Γ_1, we define a matrix $F_1^{(k)}$ ($k = 0, 1, \ldots, M - 1$) by the following complex contour integral:

$$F_1^{(k)} = \frac{1}{2\pi i} \int_{\Gamma_1} (z - \gamma_1)^k (zB - A)^{-1} B \, dz. \qquad (7.1)$$

Then, each $F_1^{(k)}$ becomes a filter, that is, a projection operator onto the eigenspace of the eigenvalues $\lambda_1, \lambda_2, \ldots, \lambda_m$. Thus, by operating each of $F_1^{(0)}, F_1^{(1)}, \ldots, F_1^{(M-1)}$ to some set of L vectors, $V = [\mathbf{v}_1, \mathbf{v}_2, \ldots, \mathbf{v}_L]$, we obtain LM vectors that belong to the eigenspace of $\lambda_1, \lambda_2, \ldots, \lambda_m$. By applying the Rayleigh–Ritz procedure [20] to these vectors, we get the eigenvalues $\lambda_1, \lambda_2, \ldots, \lambda_m$ and the corresponding eigenvectors.

---

[2]There are many variants of the Sakurai–Sugiura method. The variant explained here is called the CIRR (Contour Integral Rayleigh–Ritz) method. A method to estimate m, the number of eigenvalues in Γ_1, has also been developed [18].

Fig. 7.6 Computation of eigenvalues by the Sakurai–Sugiura method

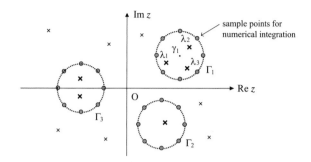

The contour integral of Eq. (7.1) is actually computed by a numerical integration formula, which has sample points on Γ_1. Since the evaluation of $(zB - A)^{-1}BV$ can be done for each sample point independently, the Sakurai–Sugiura method has large grain parallelism. In addition, if several closed curves $\Gamma_1, \Gamma_2, \ldots, \Gamma_p$ are used to enclose the eigenvalues, we have additional parallelism. The Sakurai–Sugiura and the filter diagonalization methods are expected to see widespread use in the future as methods possessing large grain parallelism. A parallel eigensolver z-Pares [48], which implements the Sakurai–Sugiura method, has already been released. For the variants of the Sakurai–Sugiura method and their mutual relationship, the readers are referred to [27].

7.5.2 Reduction of the Data Transfer

Next, we explain techniques to reduce the amount and frequency of data transfer. We first deal with techniques for dense matrix algorithms, and then those for sparse matrix algorithms.

7.5.2.1 Techniques for Dense Matrix Algorithms

Assume that the target computer has only one core and one level of cache, and we want to minimize the data transfer between the main memory and the cache. Today's computers have more than one level of cache and multiple cores and nodes, but the discussion below can be extended to such a case as well. We assume that the matrix size is n and the cache can store up to M words.

In dense numerical linear algebra, blocked (tile) algorithms have long been used. In these algorithms, the matrix is divided into blocks of size $L \times L$, where $L \equiv \sqrt{M/3}$ (Fig. 7.7). Thus, three blocks can be stored in the cache. Then we reformulate the target algorithm by treating the blocks as if they were matrix elements. In this way, many dense algorithms, such as matrix–matrix multiplication, the LU, Cholesky and

Fig. 7.7 Partitioning of a
matrix for blocked
algorithms

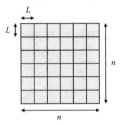

QR decompositions, orthogonal reduction to tridiagonal or Hessenberg matrices,
can be transformed into blocked ones. As an example, the algorithm of blocked
Cholesky decomposition is listed as Algorithm 1. The algorithm is almost identical
to the standard Cholesky decomposition, except that operation on the elements are
replaced by those on blocks. In the blocked algorithms, most of the computations
become matrix–matrix multiplications, which can be done within the cache due to
the definition of L. Hence, the cache can be used efficiently. Blocked algorithms
have been widely adopted in numerical linear algebra libraries such as LAPACK and
ScaLAPACK.

Recent studies show the optimality of various blocked algorithms in the sense of
minimizing data transfer. For example, Ballard et al. showed the following theorem
on the blocked Cholesky decomposition [3].

Theorem 1 *Under the setting stated above, the blocked Cholesky decomposition
with block size $L = \sqrt{M/3}$ minimizes the data transfer between the cache and the
main memory in the sense of order.*

Algorithm 1 Blocked Cholesky decomposition

1: **for** $K = 1, \ldots, n/L$ **do**
2:      $A_{KK} := \text{Cholesky}(A_{KK})$ ▷ Cholesky decomposition of a diagonal block
3:      **for** $I = K + 1, \ldots, n/L$ **do**
4:          $A_{IK} := A_{IK} A_{KK}^{-\top}$ ▷ generation of the block pivot column
5:      **end for**
6:      **for** $J = K + 1, \ldots, n/L$ **do**
7:          **for** $I = J, \ldots, n/L$ **do**
8:              $A_{IJ} := A_{IJ} - A_{IK} A_{JK}^{\top}$ ▷ matrix-matrix multiplication
9:          **end for**
10:     **end for**
11: **end for**

In the proof of this theorem, we note that the lower bound of data transfer has
already been determined for matrix–matrix multiplication, and construct an algorithm to compute matrix–matrix multiplication using Cholesky decomposition. Then,
it follows that the lower bound of data transfer of Cholesky decomposition is, up to
some multiplicative constant, larger than or equal to that of matrix–matrix multipli-

non-contiguous addresses
within a block.

contiguous addresses
within a block.

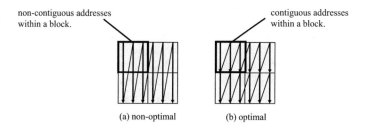

(a) non-optimal   (b) optimal

Fig. 7.8 Optimal data storage format for blocked algorithms

cation. Finally, by showing that the amount of data transfer of the blocked Cholesky decomposition attains this lower bound, we can complete the proof.

So far, we have discussed the amount of data transfer. We also need to discuss the number of data transfers. Assume that we can move only a contiguous region of the main memory to the cache by one data transfer. Under this assumption, the blocked Cholesky decomposition using the standard (column-by-column) matrix storage format cannot minimize the number of data transfers, because the L^2 elements in a block are not stored in a contiguous region in the memory (Fig. 7.8a). To solve the problem, we change the storage scheme so that the elements in a block have contiguous memory addresses (Fig. 7.8b). Then, it can be shown that the number of data transfer is also minimized. An algorithm is called *communication optimal* if it minimizes both the amount and the number of data transfer.

Lower bounds on the amount and the number of data transfer have been known for many matrix computations, including matrix–matrix multiplication (both the standard $O(n^3)$ algorithm and Strassen's algorithm), the LU, Cholesky, and QR decompositions, linear least squares and eigenvalue, and singular value decompositions [4]. Note however that this does not necessarily mean that communication optimal algorithms are known for all of these computations. The development of communication optimal algorithms, which attain both of these bounds, has been an active research area recently.

7.5.2.2 Techniques for Sparse Matrix Algorithms

As an example of algorithms for sparse matrices, we focus on the Krylov subspace methods. It is a collective name of methods that seek an approximate solution in the Krylov subspace $K_m(A; \mathbf{b}) = \text{span}\{\mathbf{b}, A\mathbf{b}, A^2\mathbf{b}, \ldots, A^{m-1}\mathbf{b}\}$ obtained by multiplying a starting vector $\mathbf{b}$ by a matrix A repeatedly. The Krylov subspace methods have applications in such diverse areas as the solution of linear simultaneous equations, eigenvalue problems and computation of matrix functions. In these algorithms, the main part of computation is sparse matrix–vector multiplications $\mathbf{y} = A\mathbf{x}$, and several inner products and norm computations of vectors are also required at each step. From the viewpoint of data transfer, there are two major problems: (i) data reuse in the sparse matrix–vector multiplication $\mathbf{y} = A\mathbf{x}$ is limited, because each matrix element is used only once and then discarded and (ii) inner products and norm computations require AllReduce operation among all nodes, which incurs large latency when the number of nodes is large. Hence, the challenge here is to enhance data reuse in the sparse matrix–vector multiplication and at the same time, reduce the number of AllReduce by putting several inner products and norm computations together. In the following, we explain the techniques to achieve these goals by taking the GMRES (Generalized Minimum RESidual) method [20] as an example.

GMRES is the most conceptually simple linear equation solver for nonsymmetric matrices based on the Krylov subspace method. In the GMRES method, we construct the sequence of Krylov subspaces $K_1(A; \mathbf{b}) \subset K_2(A; \mathbf{b}) \subset K_3(A; \mathbf{b}) \subset \cdots$ along with their orthogonal basis $\mathbf{q}_1, \mathbf{q}_2, \mathbf{q}_3, \ldots$, by starting from $K_1(A; \mathbf{b}) = \{\mathbf{b}\}$ and repeating multiplication by A and Gram–Schmidt orthogonalization. At the same time, at each step m, the approximate solution $\mathbf{x}_m \in K_m(A; \mathbf{b})$ is computed so that the residual $\|\mathbf{r}_m\|_2 = \|\mathbf{b} - A\mathbf{x}_m\|_2$ is minimized. The computationally dominant parts consist of matrix–vector multiplications by A and incremental orthogonalization to construct the orthogonal basis. These parts are depicted as Algorithm 2. In the following, we report two approaches to enhance the efficiency of these parts.

Algorithm 2 Construction of Krylov subspaces and their orthogonal basis

1: $\mathbf{q}_1 = \mathbf{b}/\|\mathbf{b}\|_2$
2: **for** $n = 1, 2, 3, \ldots$ **do**
3: $\quad \mathbf{v} = A\mathbf{q}_n$ ▷ generation of a new vector (expansion of the subspace)
4: $\quad$ **for** $j = 1, \ldots, n$ **do**
5: $\quad\quad h_{jn} = \mathbf{q}_j^\top \mathbf{v}$ ▷ orthogonalization
6: $\quad\quad \mathbf{v} := \mathbf{v} - h_{jn}\mathbf{q}_j$
7: $\quad$ **end for**
8: $\quad h_{n+1,n} = \|\mathbf{v}\|_2$ ▷ normalization
9: $\quad \mathbf{q}_{n+1} = \mathbf{v}/h_{n+1,n}$
10: **end for**

Block GMRES method [39]

In this method, we start with multiple (ℓ) initial vectors $R^{(0)} = [\mathbf{r}_1^{(0)}, \ldots, \mathbf{r}_\ell^{(0)}]$ and seek for a solution in the block Krylov subspace $K_m(A; R^{(0)})$. Consequently, the matrix–vector product $\mathbf{y} = A\mathbf{x}$ at each step is replaced with the matrix–matrix product $Y = AX$ and thus each element of A fetched from memory is used ℓ times, enhancing the data reuse of A by ℓ times. The number of AllReduce operations in each step is unchanged, although the amount of data transferred is increased by ℓ times (Table 7.1). When solving ℓ sets of linear simultaneous equations with the same coefficient matrix A and different right-hand sides using this method, the numbers of access to A and AllReduce operations for each step are the same as those of the standard GMRES method. Hence, the execution time for each step should be less than ℓ times that of the standard GMRES method. Furthermore, since the approximate solutions are sought in a larger subspace $K_m(A; R^{(0)})$, convergence should be faster. Consequently, the block GMRES method can be a very efficient method in this situation.

Table 7.1 Comparison of the standard and block GMRES methods (for one step)

	Standard GMRES	Block GMRES
Computational work (matrix-vector product)	1 ($y = A x$)	ℓ ($Y = AX$)
Number of access to A	1 ($y = A x$)	1 ($Y = AX$)
Number of AllReduce operation	1	1

Table 7.2 Comparison of the standard and k-step GMRES methods (for one step)

	Standard GMRES	k-step GMRES
Computational work (matrix-vector product)	1	$1 + \alpha$
Number of access to A	1	$1/k$
Number of AllReduce operation	1	$1/k$

Blocked algorithms have been proposed for other Krylov subspace methods such as the CG (Conjugate Gradient) method [35], QMR (Quasi-Minimum Residual) method, [15], Bi-CGSTAB (BiConjugate Gradient STABilized) method [22] and IDR(s) (Induced Dimension reduction (s)) method [11]. One difficulty with the blocked method is that the residuals computed with the recurrence formula tend to deviate from the true residuals as ℓ increases, making accurate solutions difficult to obtain [41]. To suppress this deviation, modifications of the recurrence relations have been proposed [41].

k-step GMRES method [25]

This is a variant of the GMRES method, in which k matrix–vector products $A\mathbf{r}^{(m)}$, $A^2\mathbf{r}^{(m)}, \ldots, A^k\mathbf{r}^{(m)}$ are performed at once and the Krylov subspace is expanded by k dimensions at once. If A is a matrix arising from the finite element or finite difference discretization of a partial differential equation, the operation of A is local on the mesh. Hence, if the elements of $\mathbf{r}^{(m)}$ corresponding to some region on the mesh are fetched into the cache, computation of the elements $A\mathbf{r}^{(m)}, A^2\mathbf{r}^{(m)}, \ldots, A^k\mathbf{r}^{(m)}$ corresponding to this region, except for the boundary elements, can be carried out using only the fetched elements. Thus, the data reuse of A and $\mathbf{r}^{(m)}$ can be enhanced. Moreover, orthogonalization can be performed for k vectors at once, reducing the number of AllReduce by $1/k$ (see Table 7.2).

One drawback of this method is that the convergence can be deteriorated when the basis vectors before orthogonalization, $A\mathbf{r}^{(m)}, A^2\mathbf{r}^{(m)}, \ldots, A^k\mathbf{r}^{(m)}$, are close to linearly dependent. To improve the linear independence, one can use some orthogonal polynomials $\{p_i(x)\}_{i=0}^{k}$ and compute $p_1(A)\mathbf{r}^{(m)}, p_2(A)\mathbf{r}^{(m)}, \ldots, p_k(A)\mathbf{r}^{(m)}$ instead of the natural basis using a three-term recurrence [2, 25].

In [40], various techniques to reduce data transfer are applied to the CG method, which is a Krylov subspace method for symmetric positive definite matrices, and their effectiveness is evaluated.

7.5.3 Fault Tolerance at the Algorithm Level

In this subsection, we highlight some numerical linear algebra algorithms possessing algorithm-level fault tolerance. We consider the situation where the target algorithm is executed across multiple nodes and one of the nodes returns an erroneous result during the computation, or does not returns a result at all (in which case, a timeout will occur). We want to design an algorithm which can continue execution even in such a case without breakdown. In doing so, we make a few assumptions. First, we assume that we can use highly reliable hardware (with lower error probability but higher computational cost), in addition to the standard hardware, to perform a critical part of the computation. This is because designing a fault- tolerant algorithm will become extremely difficult if we assume that an error can occur at any part of the computation. We also allow degradation of numerical accuracy or convergence speed to some extent in the presence of an error.

Before proceeding further, we observe that an algorithm which repeatedly improves a subspace by combining the results from multiple nodes are inherently suited for fault tolerance, because then even if the result from one node is missing, the effect will not be fatal. In addition, large grain parallelism will also be advantageous for fault tolerance, because the chance for an erroneous result to contaminate others is limited. In the following, we explain three algorithms having these properties.

MERAM Multiple Explicitly Restarted Arnoldi Method [14]
This is a fault-tolerant eigenvalue computation algorithm based on the Arnoldi method. The idea is simple: the algorithm repeats the following three steps until convergence (Fig. 7.9).

(1) Perform P independent Arnoldi method on P nodes using different starting vectors.
(2) Combine the Krylov subspaces generated by these nodes and construct a large subspace.
(3) Find P next initial vectors in this subspace and distribute each vector to each node.

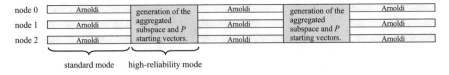

Fig. 7.9 Multiple explicitly restarted Arnoldi method

Among the three steps, step (1) is executed on standard hardware, while steps (2) and (3) are executed on highly reliable hardware. Since most of the computations are performed in step (1), the amount of computation to be performed on highly reliable hardware is relatively small. If one of the nodes fails to return the result in step (1), the number of subspaces used in step (2) decreases by one, but it will not cause a breakdown of the whole computation, although convergence can be retarded slightly. Even if one of the nodes returns a totally irrelevant subspace due to a computational error, it only means that the aggregated subspace is expanded into a wrong direction wastefully, and no breakdown will occur. Hence, we can say that MERAM realizes fault tolerance under the assumption that steps (2) and (3) are executed correctly.

Fault-Tolerant GMRES method [13]
This is a fault-tolerant linear equation solver based on the framework of the flexible GMRES method. The flexible GMRES method allows different preconditioning to be applied at each iteration step. In Fault-Tolerant GMRES method, erroneous computation is interpreted as an application of an improper preconditioning, which expands the subspace wastefully but otherwise has no harmful effects. Hence, an error does not cause breakdown even though it retards the convergence.

The Sakurai–Sugiura method
The Sakurai–Sugiura method for the eigenvalue problem extracts the invariant subspace of the desired eigenvalues from the input vectors using a filter (see Sect. 7.5.1.2). This filter is constructed from the contour integral (7.1) of the resolvent along a closed curve Γ_1 enclosing the wanted eigenvalues. In the actual algorithm, the integral is approximated by numerical integration. Now, assume that the function value at one of the sample points could not be obtained due to some error. In that case, we can reconstruct the quadrature to use only the function values at the remaining sample points. In this way, we can continue the computation, even though the accuracy can deteriorate slightly [38]. This is the simplest approach to give the Sakurai–Sugiura method fault tolerance.

Actually, there is a more sophisticated approach that does not require the reconstruction of the quadrature formula [28]. Assume that the product of the parameters L and M is sufficiently larger than the number of eigenvalues m in Γ_1. When the number of sample points is K, the LM vectors obtained by applying the M filters to $V = [\mathbf{v}_1, \mathbf{v}_2, \ldots, \mathbf{v}_L]$ are linear combinations of the KL vectors $(z_j B - A)^{-1} B \mathbf{v}_i$ $(i = 1, \ldots, L)(j = 0, \ldots, K - 1)$. Now, suppose that an error occurred in the computation at the sample point z_j. This affects L vectors among the KL vectors, so the dimension of the erroneous subspace is at most L. Hence, if the LM vectors obtained by the application of the filters are linearly independent, we can (virtually) divide the subspace $\mathscr{S}$ spanned by them into the direct sum of two subspaces, namely the L-dimensional subspace $\mathscr{S}_e$ contaminated by the error and the $(LM - L)$-dimensional subspace $\mathscr{S}_c$ not affected by the error. If $LM - L \geq m$, we can recover the eigenvalues in Γ_1 and the corresponding eigenvectors correctly, by performing the Rayleigh–Ritz procedure in $\mathscr{S}_c$. In reality, we perform the Rayleigh–Ritz procedure in $\mathscr{S} = \mathscr{S}_c \oplus \mathscr{S}_e$, but because $\mathscr{S} \supset \mathscr{S}_c$ it will produce an accurate eigenvalues and eigenvectors as would be obtained if the Rayleigh–Ritz procedure were applied to $\mathscr{S}_c$. Note that in the actual algorithm, we need to make $LM - L$ somewhat larger than m to avoid the influence from the eigenvalues outside Γ_1. The above result is interesting in that it shows that the Sakurai–Sugiura method has a built-in mechanism for fault tolerance.

So far, we have discussed fault-tolerant algorithms for sparse matrices. For dense matrices, direct methods are widely used for both linear simultaneous equations and eigenvalue problems. Since direct methods usually have no redundancy, unlike the subspace methods, any error occurring during the computation can cause disastrous results. So the design of a fault-tolerant algorithm is harder, and it would be more realistic to resort to general techniques such as checkpointing. Note however that it is possible to rollback the computation after an error occurs, if the computation process is reversible and the reverse computation can be performed stably. One example of such computation is algorithms based on orthogonal transformations, such as the Householder QR decomposition. In fact, a fault-tolerant QR decomposition algorithm for a CPU-GPU hybrid environment has been proposed [12]. In this algorithm, an error will cause an increase in the computational time due to the rollback, but will not affect the accuracy of the computed results.

7.5.4 Reducing the Order of Computational Work by Approximation Algorithms

By introducing approximation into the computation, we can reduce the order of computational work. As an example of this, we explain a stochastic algorithm for the so- called CX decomposition. We also discuss numerical methods based on low-rank approximation of tensors, which are effective for high dimensional problems.

7.5.4.1 Stochastic Algorithms for the CX Decomposition

The singular value decomposition (SVD) is known as a decomposition that gives the best low-rank approximations of a matrix. Due to this property, SVD is used in many applications such as image processing, signal processing, and information retrieval. However, when the input matrix A is m by n ($m \geq n$), the computational cost of SVD is $O(mn^2)$, which is fairly large. Consequently, in big data applications, it is difficult to use the SVD in its original form.

As an alternative to the SVD, the *CX decomposition* has attracted attention recently. In the CX decomposition of $A \in \mathbb{R}^{m \times n}$, we seek for $C \in \mathbb{R}^{m \times k}$ and $X \in \mathbb{R}^{k \times n}$ that minimize the Frobenius norm $\|A - CX\|_F$ under the condition that C is a matrix obtained by choosing k column vectors of A. This corresponds to choosing the k column vectors that best represent the feature of A. Once C has been fixed, it can be shown that the optimal choice of X is $X = C^+A$, where C^+ is the Moore–Penrose generalized inverse of C. Thus, the problem reduces to choosing C.

The CX decomposition has the advantage of being easy to interpret, because the column vectors of C are sampled from those of A. In contrast, in the case of SVD, even if the elements of A are positive (or integers), the elements of the singular vectors are in general not positive (or integers). This sometimes makes the singular vectors difficult to interpret. Thanks to this advantage and the existence of a fast approximation algorithm, the CX decomposition has rapidly gained popularity recently.

To choose the columns of C, quantities called *statistical leverage* are used [6]. Let the truncated SVD of rank k of A be $A_k = U_k \Sigma_k V_k^\top$. Then, the statistical leverage p_j of the jth column of $V_k^\top$ is defined as follows:

$$p_j = \frac{\| (V_k^\top)^{(j)} \|_2}{k},\tag{7.2}$$

where $(V_k^\top)^{(j)}$ is the jth column vector of $V_k^\top$. Now, suppose that we sample the columns of A according to the probability $\{p_j\}_{j=1}^n$. Then, it can be shown that the following inequality holds with probability larger than or equal to 0.9.

$$\|A - CC^+ A\|_\mathrm{F} \le (1 + \epsilon)\|A - A_k\|_\mathrm{F}.\tag{7.3}$$

Here, ϵ is a small quantity depending on the number of samples. Eq. (7.3) shows that with high probability, the CX decomposition obtained in this way approximates A_k, the best rank-k approximation of A, to high relative accuracy. Note however that this method of computing the CX decomposition is not practical, because it requires the truncated singular value decomposition to compute p_j.

To overcome this difficulty, a fast approximation algorithm for the statistical leverages has been developed [10]. The idea here is to transform A into another matrix A' with *uniform* statistical leverages, by pre-multiplying A with some orthogonal matrix. This becomes possible with the help of Johnson & Lindenstrauss's lemma and the fast Walsh–Hadamard transform. For the transformed matrix A', uniform sampling of the column vectors gives the CX decomposition with the stochastic error estimate (7.3). Then, we can compute the CX decomposition of A from that of A'. This technique has enabled the CX decomposition with the stochastic relative error estimate (7.3) to be computed with $O(mn \log m)$ work, leading to a great breakthrough.

7.5.4.2 Numerical Methods Based on Low-Rank Approximation of Tensors

In many body problems of quantum mechanics and transportation problems described by the Boltzmann equation, we have to solve a partial differential equation in a high-dimensional space. In such a case, if we use a regular grid and adopt the function values at each grid point as variables, the number of variables grows exponentially with the number of dimensions, making the computation infeasible. As an alternative, methods based on low-rank approximations of tensors have attracted attention recently. In these methods, we regard the set of values on the grid points as a tensor, and make use of its low-rank approximation to save memory requirement and computational work [21, 23]. As methods for low-rank approximation, the Tensor Train format or hierarchical Tucker decomposition are widely used. To solve the partial differential equation, we regard the coefficient matrix, which is obtained by discretizing the partial differential operator on the grid, also as a tensor, express it using a low-rank approximation, and use iterative methods, which require only matrix–vector products. Then, if the product of a matrix and a vector, both of which are expressed as low-rank approximations, can itself be computed efficiently as a low-rank approximation, all the computations can be done using only low-rank approximations and the problem of exponential growth of computational work is greatly mitigated. This type of approach will become important in various field of numerical computations in the future.

7.5.5 Efficient Parallel Algorithms Under Strong Scaling Conditions

In concluding this chapter, we focus on parallel numerical linear algebra algorithms that are efficient under strong scaling conditions, taking diagonalization of a symmetric matrix as an example.

Let $A \in \mathbb{R}^{n \times n}$ be a real symmetric or complex Hermitian matrix. It is well known that diagonalization of A is equivalent to computing all the eigenvalues and eigenvectors of A. Such computation is important in various areas of scientific computing, notably, quantum chemistry, and electronic structure calculation. Today, matrices as large as $n \geq 10^6$ can be diagonalized with supercomputers like the K computer. On the other hand, there is a strong need to diagonalize medium size matrices quickly. Here, we consider the problem of diagonalizing a symmetric matrix with $n = 10,000$ as quickly as possible. Such a problem arises, for example, in the molecular orbital method for quantum chemistry. In the molecular orbital method, the so-called multi-electron integrals, which generate the matrix to be diagonalized, require $O(n^4)$ or more computational work, and have a large degree of parallelism. Hence, it is quite common to parallelize this part using more than 10,000 computing nodes. In that case, the diagonalization part can become a performance bottleneck, although it requires only $O(n^3)$ work. Thus, the challenge is to parallelize the diagonalization of a medium-sized matrix efficiently on a massively parallel computer. When $n = 10,000$, the performance of ScaLAPACK eigensolver saturates around several hundred nodes and then degrades, so it cannot exploit the potential performance of tens of thousands nodes allocated for the multi-electron integrals. Hence, there is a strong need for an eigensolver with better strong scalability.

The main reason for the performance saturation and degradation of the ScaLAPACK eigensolver is a large number of internode communications occurring in the

Fig. 7.10 Time for
diagonalizing a
$10,000 \times 10,000$ real
symmetric matrix on the K
computer

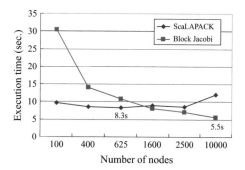

tri-diagonalization step, which is a preprocessing stage for eigenvalue computation. In fact, the performance data on the K computer for the $n = 10,000$ case reveals that more than 70% of the total computation time is devoted to internode communication and a large part of it is the startup time for communication.

To develop an eigensolver with better strong scalability, we chose to adopt the block-Jacobi method, which is a variant of the Jacobi method for the symmetric eigenvalue problem. Compared with the standard methods based on tri-diagonalization, the block-Jacobi method requires roughly ten times computational work, but smaller number of internode communications. The number of communications in the block-Jacobi method is $O(\sqrt{P} \log_2 P * Iter)$, where P is the number of nodes and $Iter$ is the number of iterations, which is much smaller than $O(n \log_2 P)$ of tri-diagonalization based methods when $n = P = 10,000$. Hence, despite the much larger computational work, the block-Jacobi method can outperform the tri-diagonalization based methods in a situation where communication overhead is dominant.

We implemented the block-Jacobi method on the K computer and compared its performance with that of ScaLAPACK, by fixing the problem size to $n = 10,000$ [30]. The result is shown in Fig. 7.10. As can be seen from the graph, the block-Jacobi-based solver has better scalability and is faster when $P = 10,000$ than the best case of ScaLAPACK. We expect to improve the performance of the block-Jacobi solver further, by incorporating preprocessing and using better pivoting strategy. Thus, it can have a performance advantage over standard solvers under the condition of strong scaling. The block- Jacobi method has proved to outperform conventional methods also for the singular value decomposition under strong scaling conditions [31].

The above example shows that the most important point in designing an efficient algorithm under strong scaling conditions is to reduce the communication and synchronization overheads. It is sometimes advantageous to pursue this objective even at the cost of increasing the computational work. Based on this principle, Nakatsukasa et al. propose a novel communication avoiding eigenvalue and singular value computation algorithms using the polar decomposition [33]. It is possible that they become a method of choice in the coming exaFLOPS era.

7.6 Conclusion

ExaFLOPS supercomputers to appear in the near future will be characterized by huge and hierarchical parallelism, widening discrepancy between the floating-point performance and data transfer speed, and increase in the failure rate. On the other hand, new requirements will emerge from the application side, such as higher parallel efficiency under strong scaling conditions and reduction of the order of computational work to solve very large problems in affordable time. As a result, the following considerations will become important in the development of numerical linear algebra algorithms in the exaFLOPS era.

- sufficient parallelism to utilize 10^9 order cores and adaptability to the hierarchical hardware.
- reduction of the amount and number of data transfer.
- fault tolerance at algorithm level.
- reduction of the order of computational work by introducing (possibly stochastic) approximation.
- efficiency under strong scaling conditions.

In this chapter, we made a survey of recent research efforts conducted along these directions. While our survey has focused only on the algorithmic side, implementation and optimization techniques to exploit the potential of exaFLOPS machines are equally important. Some of such techniques are covered in other chapters of this book.

Exercises

1. Let $A \in \mathbb{R}^{m \times n}$ be a matrix with $m \geq 2n$. Prove that the TSQR algorithm shown in Fig. 7.5 actually computes the QR decomposition $A = QR$, where $Q \in \mathbb{R}^{m \times m}$ is an orthogonal matrix and $R \in \mathbb{R}^{m \times n}$ is an upper triangular matrix. In particular, explain how to compute Q from $Q_1^{(1)}$, $Q_2^{(1)}$ and $Q^{(2)}$.
2. Consider applying the Sakurai–Sugiura method to the symmetric eigenvalue problem $A\mathbf{x} = \lambda\mathbf{x}$. Show that the matrix $F_1^{(0)}$ obtained by setting $B = I$ (the identity matrix) in Eq. (7.1) gives an orthogonal projector onto the eigenspace of A corresponding to the eigenvalues $\lambda_1, \lambda_2, \ldots, \lambda_m$ within Γ_1.
3. Implement Algorithm 1 and compare its performance with that of the standard (non-blocked) Cholesky decomposition, by varying the block size L. Discuss how the value of L that minimizes the execution time of Algorithm 1 is related to the cache size.

References

1. C. Audouze, M. Massot, S. Volz, Symplectic multi-time step parareal algorithms applied to molecular dynamics, https://hal.archives-ouvertes.fr/hal-00358459 (2009), 18 pages
2. Z. Bai, A Newton basis GMRES implementation. IMA J. Numer. Anal. **14**, 563–581 (1994)

3. G. Ballard, J. Demmel, O. Holtz, O. Schwartz, Communication-optimal parallel and sequential Cholesky decomposition. SIAM J. Sci. Comput. **32**, 3495–3523 (2010)
4. G. Ballard, J. Demmel, O. Holtz, O. Schwartz, Minimizing communication in numerical linear algebra. SIAM J. Matrix Anal. Appl. **32**, 866–901 (2011)
5. G. Bosilca, A. Bouteiller, A. Danalis, T. Herault, P. Lemarinier, J. Dongarra, DAGuE: a generic distributed DAG engine for high performance computing. Parallel Comput. **38**, 27–51 (2012)
6. C. Boutsidis, M.W. Mahoney, P. Drineas, An improved approximation algorithm for the column subset selection problem, in *SODA'09 Proceedings of the twentieth Annual ACM-SIAM Symposium on Discrete Algorithms* (2009), pp. 968–977
7. cuSPARSE, http://docs.nvidia.com/cuda/cusparse/
8. J. Demmel, L. Grigori, M. Hoemmen, J. Langou, Communication-optimal parallel and sequential QR and LU factorizations. SIAM J. Sci. Comput. **34**, A206–A239 (2012)
9. J. Dongarra, M. Faverge, T. Herault, M. Jacquelin, J. Langou, Y. Robert, Hierarchical QR factorization algorithms for multi-core cluster systems. Parallel Comput. **39**, 212–232 (2013)
10. P. Drineas, M. Magdon-Ismail, M.W. Mahoney, D.P. Woodruff, Fast approximation of matrix coherence and statistical leverage. J. Mach. Learn. Res. **13**, 3475–3506 (2012)
11. L. Du, T. Sogabe, B. Yu, Y. Yamamoto, S.-L. Zhang, A block IDR(s) method for nonsymmetric linear systems with multiple right-hand sides. J. Comput. Appl. Math. **235**, 4095–4106 (2011)
12. P. Du, P. Luszczek, S. Tomov, J. Dongarra, Soft error resilient QR factorization for hybrid system. J. Comput. Sci. **4**, 457–464 (2013)
13. J. Elliott, M. Hoemmen, F. Mueller, Evaluating the impact of SDC on the GMRES iterative solver, in *IPDPS'14 Proceedings of the 2014 IEEE 28th International Parallel and Distributed Processing Symposium* (IEEE Computer Society Press, 2014), pp. 1193–1202
14. N. Emad, S. Petiton, G. Edjlali, Multiple explicitly restarted Arnoldi method for solving large eigenproblems. SIAM J. Sci. Comput. **27**, 253–277 (2005)
15. R. Freund, M. Malhotra, A block QMR algorithm for non-Hermitian linear systems with multiple right-hand sides. Linear Algebra Appl. **254**, 119–157 (1997)
16. T. Fukaya, Y. Nakatsukasa, Y. Yanagisawa, Y. Yamamoto, CholeskyQR2: a simple and communication-avoiding algorithm for computing a tall-skinny QR factorization on a large-scale parallel system, in *Proceedings of the 5th Workshop on Latest Advances in Scalable Algorithms for Large-Scale Systems (ScalA'14)* (IEEE Press, 2014), pp. 31–38
17. T. Fukaya, T. Imamura, Y. Yamamoto, Performance analysis of the Householder-type parallel tall-skinny QR factorizations toward automatic algorithm selection. Lecture Notes in Computer Science, vol. 8969 (Springer, 2015), pp. 269–283
18. Y. Futamura, H. Tadano, T. Sakurai, Parallel stochastic estimation method of eigenvalue distribution. JSIAM Lett. **2**, 127–130 (2010)
19. E. Gallopoulos, B. Philippe, A.H. Sameh, *Parallelism in Matrix Computations* (Springer, 2016)
20. G.H. Golub, C.F. Van Loan, *Matrix Computations*, 4th edn. (John Hopkins University Press, Baltimore, 2012)
21. L. Grasedyck, D. Kressner, C. Tobler, A literature survey of low-rank tensor approximation techniques. GAMM-Mitteilungen **36**, 53–78 (2013)
22. A. El Guennouni, K. Jbilou, H. Sadok, A block version of BiCGSTAB for linear systems with multiple right-hand sides. Electron. Trans. Numer. Anal. **16**, 129–142 (2003)
23. W. Hackbusch, *Tensor Spaces and Numerical Tensor Calculus* (Springer, 2012)
24. M. Hegland, Divide and conquer for the solution of banded linear systems of equations, in *Proceedings of the Fourth Euromicro Workshop on Parallel and Distributed Processing* (IEEE Computer Society Press, 1996), pp. 394–401
25. M. Hoemmen, *Communication-Avoiding Krylov Subspace Methods* (Computer Science Division, University of California, Berkeley, 2010). Ph. D Thesis
26. HPCI Technical Roadmap White Book, http://open-supercomputer.org/wp-content/uploads/2012/03/hpci-roadmap.pdf, Mar 2012
27. A. Imakura, L. Du, T. Sakurai, Relationships among contour integral-based methods for solving generalized eigenvalue problems. Jpn. J. Ind. Appl. Math. **33**, 721–750 (2016). https://doi.org/10.1007/s13160-016-0224-x

28. A. Imakura, Y. Futamura, T. Sakura, An error resilience strategy of a complex moment-based eigensolver, in *Proceedings of International Workshop on Eigenvalue Problems: Algorithms, Software and Applications, in Petascale Computing (EPASA)* (Springer) (to appear)
29. K computer, http://www.aics.riken.jp/en/k-computer/about/
30. S. Kudo, Y. Takahashi, T. Fukaya, Y. Yamamoto, Implementation of an eigensolver based on the block Jacobi method on a massively parallel computer (in Japanese), Presented at *The Annual Meeting of the Japan Society for Industrial and Applied Mathematics* (Fukuoka, Japan, 2013)
31. S. Kudo, Y. Yamamoto, M. Bečka, M. Vajteršic, Performance analysis and optimization of the parallel one-sided block Jacobi SVD algorithm with dynamic ordering and variable blocking. Concurr. Comput. Pract. Exp. **26**, 24 pages (2017)
32. D. Mori, Y. Yamamoto, S.-L. Zhang, Backward error analysis of the AllReduce algorithm for Householder QR decomposition. Jpn. J. Ind. Appl. Math. **29**, 111–130 (2011)
33. Y. Nakatsukasa, N.J. Higham, Stable and efficient spectral divide and conquer algorithms for the symmetric eigenvalue decomposition and the SVD. SIAM J. Sci. Comput. **35**, A1325–A1349 (2013)
34. S. Nishino, T. Fujiwara, H. Yamasaki, Nanosecond quantum molecular dynamics simulations of the lithium superionic conductor $Li_{4-x}Ge_{1-x}P_xS_4$. Phys. Rev. B **90**, 024303 (2014)
35. D.P. O'Leary, The block conjugate gradient algorithm and related methods. Linear Algebra Appl. **29**, 293–322 (1980)
36. E. Polizzi, A.H. Sameh, A parallel hybrid banded system solver: the SPIKE algorithm. Parallel Comput. **32**, 177–194 (2006)
37. T. Sakurai, H. Tadano, A Rayleigh-Ritz type method with contour integral for generalized eigenvalue problems. Hokkaido Math. J. **36**, 669–918 (2007)
38. K. Shirasuna, T. Sakurai, On fault-tolerance of an eigensolver using contour integration (in Japanese), presented at *The Spring Meeting of the Japan Society for Industrial and Applied Mathematics*, Tokyo, Mar 2011
39. V. Simoncini, E. Gallopoulos, Convergence properties of block GMRES and matrix polynomials. Linear Algebra Appl. **247**, 97–119 (1996)
40. R. Suda, C. Li, D. Watanabe, Y. Kumagai, A. Fujii, T. Tanaka, Communication-avoiding CG method: new direction of Krylov subspace methods towards Exa-scale computing. RIMS Kokyuroku **1995**, 102–111 (2016), https://repository.kulib.kyoto-u.ac.jp/dspace/bitstream/2433/224703/1/1995-14.pdf
41. H. Tadano, T. Sakurai, Y. Kuramashi, A new block Krylov subspace method for computing high accuracy solutions. JSIAM Lett. **1**, 44–47 (2009)
42. S. Toledo, E. Rabani, Very large electronic structure calculations using an out-of-core filter-diagonalization method. J. Comput. Phys. **180**, 256–269 (2002)
43. H. Walker, Implementation of the GMRES method using Householder transformations. SIAM J. Sci. Stat. Comput. **9**, 152–163 (1988)
44. Y. Yamamoto, M. Igai, K. Naono, A parallel direct linear equation solver for nonsymmetric tridiagonal matrices, in *Proceedings of the 2003 SIAM Conference on Applied Linear Algebra* (SIAM, 2003), http://www.siam.org/meetings/la03/proceedings/yamamoty.PDF
45. Y. Yamamoto, Y. Hirota, A parallel algorithm for incremental orthogonalization based on the compact WY representation. JSIAM Lett. **3**, 89–92 (2011)
46. Y. Yamamoto, Y. Nakatsukasa, Y. Yanagisawa, T. Fukaya, Roundoff error analysis of the CholeskyQR2 algorithm. Electron. Trans. Numer. Anal. **44**, 306–326 (2015)
47. T. Fukaya, R. Kannan, Y. Nakatsukasa, Y. Yamamoto, Y. Yanagisawa, Shifted CholeskyQR for computing the QR factorization of ill-conditioned matrices. arXiv:1809.11085
48. z-Pares, http://zpares.cs.tsukuba.ac.jp/

Chapter 8
Fast Fourier Transform in Large-Scale Systems

Daisuke Takahashi

Abstract The fast Fourier transform (FFT) is an efficient implementation of the discrete Fourier transform (DFT). The FFT is widely used in numerous applications in engineering, science, and mathematics. This chapter presents an introduction to the basis of the FFT and its implementation in parallel computing. Parallel computation is becoming indispensable in solving the large-scale problems that arise in a wide variety of applications. The chapter provides a thorough and detailed explanation of FFT for parallel computers. The algorithms are presented in pseudocode, and a complexity analysis is provided. This chapter also provides up-to-date computational techniques relevant to the FFT in state-of-the-art processors.

Keywords Fast Fourier transform · Distributed memory parallel computer · All-to-all communication

8.1 Introduction

The fast Fourier transform (FFT) [6] is a fast algorithm for computing the discrete Fourier transform (DFT).
    Examples of FFT applications in the field of science are the following:

- Solving partial differential equations,
- Convolution and correlation calculations, and
- Density functional theory in first-principles calculations.

Examples of FFT applications in the field of engineering are the following:

- Spectrum analyzers,
- Image processing, for example, in CT scanning and MRI, and

D. Takahashi (✉)
Center for Computational Sciences, University of Tsukuba, 1-1-1 Tennodai, Tsukuba,
Ibaraki 305-8577, Japan
e-mail: daisuke@cs.tsukuba.ac.jp

© Springer Nature Singapore Pte Ltd. 2019   137
M. Geshi (ed.), *The Art of High Performance Computing for Computational
Science, Vol. 1*, https://doi.org/10.1007/978-981-13-6194-4_8

- Modulation and demodulation processing in orthogonal frequency multiplex modulation (OFDM) used in terrestrial digital television broadcasting and wireless LAN.

In this chapter, the optimization of the FFT in large-scale systems is described. First, we explain the definition of the DFT and then describe the basic idea of the FFT. Furthermore, we show that cache blocking can also be applied to the FFT. We then describe the parallel three-dimensional FFT algorithm using two-dimensional decomposition. Finally, parallel one-dimensional FFT in the GPU cluster is described.

8.2 Fast Fourier Transform

8.2.1 Discrete Fourier Transform

The DFT is given by

$$y(k) = \sum_{j=0}^{n-1} x(j)\omega_n^{jk}, \quad 0 \le k \le n-1. \tag{8.1}$$

Moreover, the inverse DFT is given by

$$x(j) = \frac{1}{n} \sum_{k=0}^{n-1} y(k)\omega_n^{-jk}, \quad 0 \le j \le n-1, \tag{8.2}$$

where $\omega_n = e^{-2\pi i/n}$ and $i = \sqrt{-1}$.

8.2.2 Basic Idea of FFT

In Eq. (8.1), for example, when $n = 4$, the DFT can be calculated as follows:

$$\begin{aligned}
y(0) &= x(0)\omega^0 + x(1)\omega^0 + x(2)\omega^0 + x(3)\omega^0, \\
y(1) &= x(0)\omega^0 + x(1)\omega^1 + x(2)\omega^2 + x(3)\omega^3, \\
y(2) &= x(0)\omega^0 + x(1)\omega^2 + x(2)\omega^4 + x(3)\omega^6, \\
y(3) &= x(0)\omega^0 + x(1)\omega^3 + x(2)\omega^6 + x(3)\omega^9.
\end{aligned} \tag{8.3}$$

Equation (8.3) can be expressed more simply in the form of a matrix–vector product using a matrix as follows:

$$
\begin{bmatrix} y(0) \\ y(1) \\ y(2) \\ y(3) \end{bmatrix} = \begin{bmatrix} \omega^0 & \omega^0 & \omega^0 & \omega^0 \\ \omega^0 & \omega^1 & \omega^2 & \omega^3 \\ \omega^0 & \omega^2 & \omega^4 & \omega^6 \\ \omega^0 & \omega^3 & \omega^6 & \omega^9 \end{bmatrix} \begin{bmatrix} x(0) \\ x(1) \\ x(2) \\ x(3) \end{bmatrix}. \tag{8.4}
$$

In Eq. (8.1), $x(j)$, $y(k)$, and ω are complex values. Therefore, in order to calculate the matrix–vector product of Eq. (8.4), n^2 complex multiplications and $n(n-1)$ complex additions are required.

When we use the relation $\omega_n^{jk} = \omega_n^{jk \bmod n}$, Eq. (8.4) can be written as follows:

$$
\begin{bmatrix} y(0) \\ y(1) \\ y(2) \\ y(3) \end{bmatrix} = \begin{bmatrix} 1 & 1 & 1 & 1 \\ 1 & \omega^1 & \omega^2 & \omega^3 \\ 1 & \omega^2 & \omega^0 & \omega^2 \\ 1 & \omega^3 & \omega^1 & \omega^1 \end{bmatrix} \begin{bmatrix} x(0) \\ x(1) \\ x(2) \\ x(3) \end{bmatrix}. \tag{8.5}
$$

The following decomposition of the matrix allows the number of complex multiplications to be reduced:

$$
\begin{bmatrix} y(0) \\ y(2) \\ y(1) \\ y(3) \end{bmatrix} = \begin{bmatrix} 1 & \omega^0 & 0 & 0 \\ 1 & \omega^2 & 0 & 0 \\ 0 & 0 & 1 & \omega^1 \\ 0 & 0 & 1 & \omega^3 \end{bmatrix} \begin{bmatrix} 1 & 0 & \omega^0 & 0 \\ 0 & 1 & 0 & \omega^0 \\ 1 & 0 & \omega^2 & 0 \\ 0 & 1 & 0 & \omega^2 \end{bmatrix} \begin{bmatrix} x(0) \\ x(1) \\ x(2) \\ x(3) \end{bmatrix}. \tag{8.6}
$$

Performing this decomposition recursively, the number of arithmetic operations can be reduced to $O(n \log n)$.

Let us consider a generalization of the decomposition. When n is divisible by two in Eq. (8.1), by decomposing the n-point data into the first half of the $n/2$-point data and the second half of the $n/2$-point data, the n-point DFT can be expressed as follows:

$$
\begin{aligned}
y(k) &= \sum_{j=0}^{n/2-1} \{ x(j)\omega_n^{jk} + x(j+n/2)\omega_n^{(j+n/2)k} \} \\
&= \sum_{j=0}^{n/2-1} \{ x(j) + x(j+n/2)\omega_n^{(n/2)k} \} \omega_n^{jk}, \quad 0 \le k \le n-1.
\end{aligned} \tag{8.7}
$$

Here, $\omega_n^{(n/2)k}$ in Eq. (8.7) can be expressed in terms of $\omega_n = e^{-2\pi i/n}$ as follows:

$$
\omega_n^{(n/2)k} = e^{-\pi i k} = \begin{cases} 1 & \text{if } k \text{ is even} \\ -1 & \text{if } k \text{ is odd} \end{cases}. \tag{8.8}
$$

Therefore, if k is divided into even and odd numbers, the n-point DFT can be decomposed into two $n/2$-point DFTs as follows:

$$y(2k) = \sum_{j=0}^{n/2-1} \{x(j) + x(j+n/2)\}\omega_n^{2jk}$$

$$= \sum_{j=0}^{n/2-1} \{x(j) + x(j+n/2)\}\omega_{n/2}^{jk}, \quad 0 \le k \le n/2 - 1, \qquad (8.9)$$

$$y(2k+1) = \sum_{j=0}^{n/2-1} \{x(j) - x(j+n/2)\}\omega_n^{j(2k+1)}$$

$$= \sum_{j=0}^{n/2-1} \left\{\{x(j) - x(j+n/2)\}\omega_n^j\right\} \omega_{n/2}^{jk}, \quad 0 \le k \le n/2 - 1. (8.10)$$

The $n/2$-point DFT in Eqs. (8.9) and (8.10) is calculated by $n^2/4$ complex multiplications and $(n/2)(n/2 - 1)$ complex additions. By this decomposition, the arithmetic operations are reduced to approximately $1/2$. Furthermore, when n is a power of two, by recursively performing this decomposition, the n-point DFT can finally be reduced to a two-point DFT and the arithmetic operations can be reduced to $O(n \log n)$.

There are two methods for FFT decomposition, namely, the decimation-in-time method and the decimation-in-frequency method. In the decimation-in-time method, n-point data are decomposed into the even-numbered $n/2$-point data and the odd-numbered $n/2$-point data, whereas in the decimation-in-frequency method, n-point data are divided into the first half of the $n/2$-point data and the second half of the $n/2$-point data, as shown in Eq. (8.7). Both methods require the same number of arithmetic operations.

Listing 1 Decimation-in-frequency FFT routine by recursive call

```
      recursive subroutine fft(x,temp,n)
      implicit real*8 (a-h,o-z)
      complex*16 x(*),temp(*)
!
      if (n .le. 1) return
!
      pi=4.0d0*atan(1.0d0)
      px=-2.0d0*pi/dble(n)
!
      do j=1,n/2
        w=px*dble(j-1)
        temp(j)=x(j)+x(j+n/2)
        temp(j+n/2)=(x(j)-x(j+n/2))*dcmplx(cos(w),sin(w))
      end do
!
      call fft(temp,x,n/2)
      call fft(temp(n/2+1),x,n/2)
```

```
!
      do j=1,n/2
        x(2*j-1)=temp(j)
        x(2*j)=temp(j+n/2)
      end do
      return
      end
```

By straightforwardly coding the concept behind Eqs. (8.9) and (8.10), we can write the decimation-in-frequency FFT routine by recursive call, as shown in Listing 1. In this routine, $\log_2 n$ recursive calls are made. Since $n/2$ complex multiplications and n complex additions and subtractions are performed on each call, the number of arithmetic operations is $O(n \log n)$. In this routine, the values of trigonometric functions are calculated by calling the functions `sin` and `cos` each time in the innermost loop, but these values are calculated in advance and are stored in the table for speeding up. In order to obtain the inverse FFT, we first invert the sign of `w=px*dble(j-1)` and calculate `w=-px*dble(j-1)` and then multiply the calculation result by $1/n$.

Listing 2 Cooley-Tukey algorithm

```
      subroutine fft(x,n,m)
!  Number of data n = 2**m
      implicit real*8 (a-h,o-z)
      complex*16 x(*),temp
!
      pi=4.0d0*atan(1.0d0)
!
      l=n
      do k=1,m
        px=-2.0d0*pi/dble(l)
        l=l/2
        do j=1,l
          w=px*dble(j-1)
          do i=j,n,l*2
            temp=x(i)-x(i+1)
            x(i)=x(i)+x(i+1)
            x(i+1)=temp*dcmplx(cos(w),sin(w))
          end do
        end do
      end do
!  Rearrange output data in bit-reversed order
      j=1
      do i=1,n-1
        if (i .lt. j) then
          temp=x(i)
          x(i)=x(j)
          x(j)=temp
        end if
        k=n/2
10      if (k .lt. j) then
          j=j-k
          k=k/2
```

```
        go to 10
      end if
      j = j + k
    end do
    return
    end
```

8.2.3 Cooley–Tukey Algorithm

By rewriting the recursive call in the FFT routine of Listing 1 into a loop, the FFT
algorithm of Listing 2, which is known as the Cooley–Tukey algorithm [6], is derived.
Note that the input data is overwritten with the output data (in-place). Figure 8.1
shows the data flow graph of a decimation-in-frequency Cooley–Tukey FFT. As
shown in Fig. 8.1, the operation in each stage is composed of two data operations
and does not affect other operations.

These two data operations are calculated as shown in Fig. 8.2, and the calculations
can be written as follows:

$$X = x + y,$$
$$Y = (x - y)\omega^j. \tag{8.11}$$

This equation is the basic operation of decimation-in-frequency or butterfly
operation.

Fig. 8.1 Data flow graph of a decimation-in-frequency Cooley–Tukey FFT ($n = 8$)

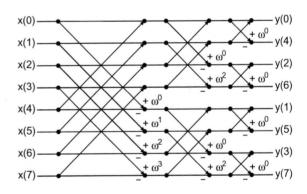

Fig. 8.2 Basic operation of decimation-in-frequency

$$X = x + y$$
$$Y = (x - y)\omega^j$$

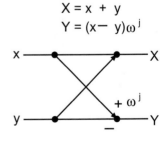

8.2.4   Bit-Reversal Permutation

In the Cooley–Tukey algorithm, the input data is overwritten with the output data, and, as shown in Fig. 8.1, and the order of the output data is by bit reversal. Table 8.1 shows the bit-reversal permutation for $n = 8$.

In order to make the order of the output data the same as the input data, it is necessary to rearrange the order obtained by bit reversal. Cache misses frequently occur because this permutation is not continuous access and the spatial locality is low.

Furthermore, the innermost loop do i=j,n,l*2, which computes the FFT, is a stride access of power of two. In order to obtain high performance, it is necessary to perform computation using as much as possible the data of one line (often 64 bytes or 128 bytes) loaded from the cache memory. However, in the case of a stride access of power of two, a situation occurs whereby only one data point among the data of one line can be used. Therefore, when considering hierarchical memory, the Cooley–Tukey algorithm is not necessarily a preferable algorithm.

The FFT routine based on the recursive call shown in Listing 1 has the same effect as hierarchically cache blocking, so the performance often increases.

8.2.5   Stockham Algorithm

The Cooley–Tukey algorithm is an in-place algorithm, in which the input data is overwritten with the output data, but it is also possible to construct an out-of-place algorithm that stores input data and output data in separate arrays. The Stockham algorithm [5], which is known as an out-of-place algorithm, is shown in Fig. 8.3.

With the Stockham algorithm, since the input data cannot be overwritten with the output data, the required memory capacity is doubled as compared with the Cooley–Tukey algorithm. However, the innermost loop is continuous access and bit-reversal permutation is unnecessary. Therefore, the Stockham algorithm is more advantageous than the Cooley–Tukey algorithm from the viewpoint of adaptability to the hierarchical memory.

Table 8.1 Bit-reversal permutation for $n = 8$

Input order	Binary notation	Bit reversal	Output order
0	000	000	0
1	001	100	4
2	010	010	2
3	011	110	6
4	100	001	1
5	101	101	5
6	110	011	3
7	111	111	7

Fig. 8.3 Stockham
algorithm

$n = 2^p, X_0(j) = x(j), 0 \le j < n$, and $\omega_n = e^{-2\pi i/n}$
$l = n/2; \ m = 1$
do $t = 1, p$
  do $j = 0, l - 1$
    do $k = 0, m - 1$
      $c_0 = X_{t-1}(k + jm)$
      $c_1 = X_{t-1}(k + jm + lm)$
      $X_t(k + 2jm) = c_0 + c_1$
      $X_t(k + 2jm + m) = \omega_{2l}^j(c_0 - c_1)$
    end do
  end do
  $l = l/2; \ m = m * 2$
end do

8.2.6 FFT Algorithm for Arbitrary Number of Data Points

In the FFT algorithm up to Sect. 8.2.5, we have assumed that the number of data in the
discrete Fourier transform n is a power of two. In this subsection, we explain the FFT
algorithm when this assumption is excluded. In order to derive the FFT algorithm
for an arbitrary number of data points, let us first consider the case of $n = n_1 n_2$.

If n has factors n_1 and n_2 ($n = n_1 \times n_2$), then indices j and k in Eq. (8.1) can be
expressed as follows:

$$j = j_1 + j_2 n_1, \quad k = k_2 + k_1 n_2. \tag{8.12}$$

We can define x and y in Eq. (8.1) as two-dimensional arrays (in column-major
order) as given below:

$$x(j) = x(j_1, j_2), \quad 0 \le j_1 \le n_1 - 1, \quad 0 \le j_2 \le n_2 - 1, \tag{8.13}$$
$$y(k) = y(k_2, k_1), \quad 0 \le k_1 \le n_1 - 1, \quad 0 \le k_2 \le n_2 - 1. \tag{8.14}$$

Substituting the indices j and k in Eq. (8.1) with those in Eq. (8.12) and using the
relation $n = n_1 \times n_2$, we derive the following equation:

$$
\begin{aligned}
y(k_2, k_1) &= \sum_{j_1=0}^{n_1-1} \sum_{j_2=0}^{n_2-1} x(j_1, j_2) \omega_n^{(j_1+j_2 n_1)(k_2+k_1 n_2)} \\
&= \sum_{j_1=0}^{n_1-1} \sum_{j_2=0}^{n_2-1} x(j_1, j_2) \omega_n^{j_1 k_2} \omega_n^{j_1 k_1 n_2} \omega_n^{j_2 k_2 n_1} \omega_n^{j_2 k_1 n_1 n_2} \\
&= \sum_{j_1=0}^{n_1-1} \left[\sum_{j_2=0}^{n_2-1} x(j_1, j_2) \omega_{n_2}^{j_2 k_2} \omega_{n_1 n_2}^{j_1 k_2} \right] \omega_{n_1}^{j_1 k_1}.
\end{aligned}
\tag{8.15}
$$

Here, the relation $\omega_n^{n_1 n_2} = \omega_n^n = 1$ is used.

Equation (8.15) shows that the n-point DFT can be decomposed into an n_1-point DFT and an n_2-point DFT. In other words, even if n is not a power of two, the number of arithmetic operations can be reduced. If we decompose the n-point DFT into an $n_1 = n/2$-point DFT and an $n_2 = 2$-point DFT recursively, the decimation-in-frequency Cooley–Tukey algorithm is derived. Moreover, if we decompose the n-point DFT into an $n_1 = 2$-point DFT and an $n_2 = n/2$-point DFT recursively, the decimation-in-time Cooley–Tukey algorithm is derived.

The number of data points to which the DFT is finally reduced in the FFT decomposition is called the radix. In Sect. 8.2.2, we explained the case in which the radix is 2, but the FFT algorithm can be constructed in the same manner, even if the radix is 3, 5, or 7. It is also possible to construct a mixed-radix FFT algorithm such that the radices are 2 and 4 [18, 23].

8.2.7 Six-Step FFT Algorithm

In this subsection, we explain the six-step FFT algorithm [4, 24] that can effectively use the cache memory. In the six-step FFT algorithm, it is possible to reduce cache misses compared to the Stockham algorithm by computing a one-dimensional FFT represented in the two-dimensional formulation described in Eq. (8.15).

The following six-step FFT algorithm [4, 24] is derived from Eq. (8.15):

Step 1: Transposition

$$x_1(j_2, j_1) = x(j_1, j_2).$$

Step 2: n_1 individual n_2-point multicolumn FFTs

$$x_2(k_2, j_1) = \sum_{j_2=0}^{n_2-1} x_1(j_2, j_1)\omega_{n_2}^{j_2 k_2}.$$

Step 3: Twiddle factor multiplication

$$x_3(k_2, j_1) = x_2(k_2, j_1)\omega_{n_1 n_2}^{j_1 k_2}.$$

Step 4: Transposition

$$x_4(j_1, k_2) = x_3(k_2, j_1).$$

Step 5: n_2 individual n_1-point multicolumn FFTs

$$x_5(k_1, k_2) = \sum_{j_1=0}^{n_1-1} x_4(j_1, k_2)\omega_{n_1}^{j_1 k_1}.$$

Step 6: Transposition
$$y(k_2,\ k_1) = x_5(k_1,\ k_2).$$

In Step 3, $\omega_{n_1n_2}^{j_1k_2} = e^{-2\pi i j_1 k_2/(n_1 n_2)}$ is a root of unity called a twiddle factor and is a complex number. Moreover, the transposition of the matrix in Steps 1 and 4 is carried out in order to make the memory access of the multicolumn FFT in Steps 2 and 5 continuous. Furthermore, transposing the matrix of Step 6 is necessary in order to make the order of the input data and the order of the output data the same.

The features of the six-step FFT algorithm are as follows:

- When $n_1 = n_2 = \sqrt{n}$, $\sqrt{n}$ individual $\sqrt{n}$-point multicolumn FFTs [24] are performed in Steps 2 and 5. The $\sqrt{n}$-point multicolumn FFTs have high locality of memory access and are suitable for a processor equipped with a cache memory.
- It is necessary to transpose the matrix three times. These three matrix transpositions are a bottleneck in the processor equipped with the cache memory.

Cache misses can be reduced by cache blocking the matrix transposition in Steps 1, 4, and 6. However, even when cache blocking is performed, the multicolumn FFT and the matrix transposition are separated. Therefore, there is a problem in that the data placed in the cache memory in the multicolumn FFT cannot be effectively reused when transposing the matrix.

8.2.8 Blocked Six-Step FFT Algorithm

In this subsection, in order to further effectively reuse the data in the cache and reduce the number of cache misses, we explain a blocked six-step FFT algorithm [19] that combines the separated multicolumn FFT and the matrix transposition in the six-step FFT. In the six-step FFT described in Sect. 8.2.7, let $n = n_1 n_2$ and let n_b be the block size. Here, it is assumed that the processor is equipped with a multilevel cache memory. The blocked six-step FFT algorithm is as follows:

1. Assume that the input data is contained in a complex array x of size $n_1 \times n_2$. At this time, while transferring n_b rows of data from $n_1 \times n_2$ array x, transfer the data to a work array work of size $n_2 \times n_b$. Here, the block size n_b is set such that the array work fits into the L2 cache.
2. Perform n_b individual n_2-point multicolumn FFTs on the $n_2 \times n_b$ array in the L2 cache. Here, it is assumed that each column FFT can be performed substantially in the L2 cache.
3. After multicolumn FFTs, multiply each element of the $n_2 \times n_b$ array work remaining in the L2 cache by the twiddle factor w. The data of this $n_2 \times n_b$ array work is then stored again in the same place as the original $n_1 \times n_2$ array x while transposing n_b rows at a time.
4. Perform n_2 individual n_1-point multicolumn FFTs on the $n_1 \times n_2$ array. Here, each column FFT can almost be performed in the L1 cache.

5. Finally, this $n_1 \times n_2$ array x is transposed by n_b rows and is stored in the $n_2 \times n_1$ array y.

Listing 3 Blocked six-step FFT algorithm

```
        complex*16  x(n1,n2),y(n2,n1),w(n1,n2),work(n2+np,nb)
!
        do ii=1,n1,nb
! Step  1:  Blocked  transposition
           do jj=1,n2,nb
              do i=ii,min(ii+nb-1,n1)
                 do j=jj,min(jj+nb-1,n2)
                    work(j,i-ii+1)=x(i,j)
                 end do
              end do
           end do
! Step  2:  n1  individual  n2-point  multicolumn
FFTs
           do i=ii,min(ii+nb-1,n1)
              call fft(work(1,i-ii+1),n2)
           end do
! Steps  3-4:  Blocked  twiddle  factor  multiplication  and
!                  transposition
           do j=1,n2
              do i=ii,min(ii+nb-1,n1)
                 x(i,j)=work(j,i-ii+1)*w(i,j)
              end do
           end do
        end do
        do jj=1,n2,nb
! Step  5:  n2  individual  n1-point  multicolumn  FFTs
           do j=jj,min(jj+nb-1,n2)
              call fft(x(1,j),n1)
           end do
! Step  6:  Blocked  transposition
           do i=1,n1
              do j=jj,min(jj+nb-1,n2)
                 y(j,i)=x(i,j)
              end do
           end do
        end do
```

The program of the blocked six-step FFT algorithm is shown in Listing 3. Here, the parameters nb and np are the blocking parameter and the padding parameter, respectively. The array work is the work array. Figure 8.4 shows the memory layout of the blocked six-step FFT algorithm. In Fig. 8.4, numbers 1 through 8 in the arrays x, work, and y indicate the sequence of accessing the array. By padding the work array work, it is possible to minimize the occurrence of cache line conflict when transferring data from array work to array x or to perform multicolumn FFTs on array work whenever possible. Note that this algorithm is a *two-pass* algorithm [4, 24]. Here, the *two-pass* algorithm reads and writes each element of the array twice. In other words, in the blocked six-step FFT algorithm, the number of arithmetic

1. Partial transposition

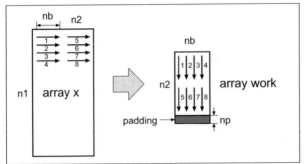

2. nb individual n2-point FFTs

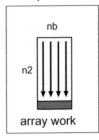

3. Partial transposition

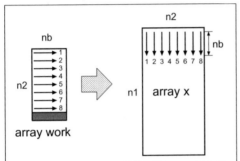

4. nb individual n1-point FFTs

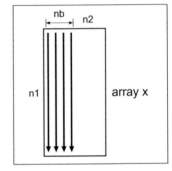

Fig. 8.4 Memory layout of the blocked six-step FFT algorithm

operations of the n-point FFT is $O(n \log n)$, whereas the number of accesses to the main memory is ideally $O(n)$.

In this subsection, it is assumed that each column FFT in Steps 2 and 4 fits into the L1 cache. However, when the problem size n is very large, it is expected that each column FFT may not fit into the L1 cache. In such a case, it is possible to calculate each column FFT in the L1 cache by reducing the problem size of each column FFT using multidimensional formulation instead of the two-dimensional formulation. However, when using a multidimensional formulation of more than three dimensions, a *two-pass* algorithm is not to be used. For example, in the case of using a three-dimensional representation, a *three-pass* algorithm is used. Thus, as the number of dimensions of the multidimensional formulation increases, FFTs of larger problem sizes can be performed. On the other hand, the number of accesses to main memory increases, indicating that the performance also depends on the capacity of the cache memory in the blocked six-step FFT.

Even if we use an out-of-place algorithm (e.g., Stockham algorithm) for the multicolumn FFTs of Steps 2 and 4, the additional array size is only $O(\sqrt{n})$. Moreover, if the output of the one-dimensional FFT is the transposed output, the transposition of the matrix of Step 5 can be omitted. In this case, the size of the array work needs only be $O(\sqrt{n})$.

8.3 Multidimensional FFTs

In this section, two-dimensional FFT algorithm and three-dimensional FFT algorithm are explained as examples of multidimensional FFT algorithms. As multidimensional FFT algorithms, there are a row–column algorithm and a vector-radix FFT algorithm [17]. We describe multidimensional FFT algorithms based on the row–column algorithm.

8.3.1 Two-Dimensional FFT Algorithm

The two-dimensional DFT is given by

$$y(k_1, k_2) = \sum_{j_1=0}^{n_1-1} \sum_{j_2=0}^{n_2-1} x(j_1, j_2)\omega_{n_1}^{j_1 k_1} \omega_{n_2}^{j_2 k_2},$$

$$0 \le k_1 \le n_1 - 1, \quad 0 \le k_2 \le n_2 - 1. \tag{8.16}$$

Moreover, the inverse two-dimensional DFT is given by

$$x(j_1, j_2) = \frac{1}{n_1 n_2} \sum_{k_1=0}^{n_1-1} \sum_{k_2=0}^{n_2-1} y(k_1, k_2)\omega_{n_1}^{-j_1 k_1} \omega_{n_2}^{-j_2 k_2},$$

$$0 \le j_1 \le n_1 - 1, \quad 0 \le j_2 \le n_2 - 1, \tag{8.17}$$

where $\omega_{n_r} = e^{-2\pi i/n_r}$ $(1 \le r \le 2)$ and $i = \sqrt{-1}$.

Equation (8.16) can be modified as follows:

$$y(k_1, k_2) = \sum_{j_2=0}^{n_2-1} \left[\sum_{j_1=0}^{n_1-1} x(j_1, j_2)\omega_{n_1}^{j_1 k_1} \right] \omega_{n_2}^{j_2 k_2}. \tag{8.18}$$

Furthermore, Eq. (8.18) can be calculated by applying the one-dimensional DFT twice as follows:

$$t(k_1, j_2) = \sum_{j_1=0}^{n_1-1} x(j_1, j_2)\omega_{n_1}^{j_1 k_1}, \tag{8.19}$$

$$y(k_1, k_2) = \sum_{j_2=0}^{n_2-1} t(k_1, j_2)\omega_{n_2}^{j_2 k_2}. \tag{8.20}$$

In other words, first, n_1-point DFTs are performed n_2 times, and n_2-point DFTs are then performed n_1 times. This method is referred to as the row–column algorithm.

The following two-dimensional FFT based on the row–column algorithm is derived from Eq. (8.18):

Step 1: n_2 individual n_1-point multicolumn FFTs

$$x_1(k_1,\ j_2) = \sum_{j_1=0}^{n_1-1} x(j_1,\ j_2)\omega_{n_1}^{j_1k_1}.$$

Step 2: Transposition

$$x_2(j_2,\ k_1) = x_1(k_1,\ j_2).$$

Step 3: n_1 individual n_2-point multicolumn FFTs

$$x_3(k_2,\ k_1) = \sum_{j_2=0}^{n_2-1} x_2(j_2,\ k_1)\omega_{n_2}^{j_2k_2}.$$

Step 4: Transposition

$$y(k_1,\ k_2) = x_3(k_2,\ k_1).$$

Listing 4 Blocked two-dimensional FFT based on row-column algorithm

```
        subroutine fft2d(x,n1,n2)
        implicit real*8 (a-h,o-z)
        complex*16 x(n1,n2),work(n2+np,nb)
! Step 1: n2 individual n1-point multicolumn FFTs
        do j=1,n2
          call fft(x(1,j),n1)
        end do
! Step 2: Blocked transposition
        do ii=1,n1,nb
          do jj=1,n2,nb
            do i=ii,min(ii+nb-1,n1)
              do j=jj,min(jj+nb-1,n2)
                work(j,i-ii+1)=x(i,j)
              end do
            end do
          end do
! Step 3: n1 individual n2-point multicolumn FFTs
          do i=ii,min(ii+nb-1,n1)
            call fft(work(1,i-ii+1),n2)
          end do
! Step 4: Blocked transposition
          do j=1,n2
            do i=ii,min(ii+nb-1,n1)
              x(i,j)=work(j,i-ii+1)
            end do
          end do
        end do
        return
        end
```

8.3.2 Blocked Two-Dimensional FFT Algorithm

In the two-dimensional FFT based on the row–column algorithm, it is necessary to transpose the matrix twice, but the algorithm can be blocked in the same manner as the blocked six-step FFT. Listing 4 shows a blocked two-dimensional FFT algorithm based on the row–column algorithm.

Here, the parameters nb and np are the blocking parameter and padding parameter, respectively. The array work is the work array. In the blocked two-dimensional FFT algorithm, the array work of size $O(n_2)$ is sufficient for the work array. In other words, the input data is overwritten with the output data, which is an in-place algorithm.

8.3.3 Three-Dimensional FFT Algorithm

The three-dimensional DFT is given by

$$y(k_1, k_2, k_3) = \sum_{j_1=0}^{n_1-1} \sum_{j_2=0}^{n_2-1} \sum_{j_3=0}^{n_3-1} x(j_1, j_2, j_3)\omega_{n_1}^{j_1 k_1} \omega_{n_2}^{j_2 k_2} \omega_{n_3}^{j_3 k_3},$$
$$0 \le k_1 \le n_1 - 1, \quad 0 \le k_2 \le n_2 - 1, \quad 0 \le k_3 \le n_3 - 1. \ (8.21)$$

Moreover, the inverse three-dimensional DFT is given by

$$x(j_1, j_2, j_3) = \frac{1}{n_1 n_2 n_3} \sum_{k_1=0}^{n_1-1} \sum_{k_2=0}^{n_2-1} \sum_{k_3=0}^{n_3-1} y(k_1, k_2, k_3)\omega_{n_1}^{-j_1 k_1} \omega_{n_2}^{-j_2 k_2} \omega_{n_3}^{-j_3 k_3},$$
$$0 \le j_1 \le n_1 - 1, \quad 0 \le j_2 \le n_2 - 1, \quad 0 \le j_3 \le n_3 - 1, \ (8.22)$$

where $\omega_{n_r} = e^{-2\pi i/n_r}$ $(1 \le r \le 3)$ and $i = \sqrt{-1}$.

The following three-dimensional FFT is derived from Eq. (8.21):

Step 1: $n_2 n_3$ individual n_1-point multicolumn FFTs

$$x_1(k_1, j_2, j_3) = \sum_{j_1=0}^{n_1-1} x(j_1, j_2, j_3)\omega_{n_1}^{j_1 k_1}.$$

Step 2: Transposition

$$x_2(j_2, j_3, k_1) = x_1(k_1, j_2, j_3).$$

Step 3: $n_3 n_1$ individual n_2-point multicolumn FFTs

$$x_3(k_2, \ j_3, \ k_1) = \sum_{j_2=0}^{n_2-1} x_2(j_2, \ j_3, \ k_1)\omega_{n_2}^{j_2 k_2}.$$

Step 4: Transposition

$$x_4(j_3, \ k_1, \ k_2) = x_3(k_2, \ j_3, \ k_1).$$

Step 5: $n_1 n_2$ individual n_3-point multicolumn FFTs

$$x_5(k_3, \ k_1, \ k_2) = \sum_{j_3=0}^{n_3-1} x_4(j_3, \ k_1, \ k_2)\omega_{n_3}^{j_3 k_3}.$$

Step 6: Transposition

$$y(k_1, \ k_2, \ k_3) = x_5(k_3, \ k_1, \ k_2).$$

Listing 5 Blocked three-dimensional FFT based on row-column algorithm

```
      subroutine fft3d(x,n1,n2,n3)
      implicit real*8 (a-h,o-z)
      complex*16 x(n1,n2,n3),ywork(n2+np,nb),zwork(n3+np,nb)
! Step 1: n2*n3 individual n1-point multicolumn FFTs
      do k=1,n3
        do j=1,n2
          call fft(x(1,j,k),n1)
        end do
! Step 2: Blocked transposition
        do ii=1,n1,nb
          do i=ii,min(ii+nb-1,n1)
            do j=1,n2
              ywork(j,i-ii+1)=x(i,j,k)
            end do
          end do
! Step 3: n3*n1 individual n2-point multicolumn FFTs
          do i=ii,min(ii+nb-1,n1)
            call fft(ywork(1,i-ii+1),n2)
          end do
! Step 4: Blocked transposition
          do j=1,n2
            do i=ii,min(ii+nb-1,n1)
              x(i,j,k)=ywork(j,i-ii+1)
            end do
          end do
        end do
      end do
      do j=1,n2
        do ii=1,n1,nb
          do i=ii,min(ii+nb-1,n1)
            do k=1,n3
              zwork(k,i-ii+1)=x(i,j,k)
            end do
          end do
! Step 5: n1*n2 individual n3-point multicolumn FFTs
```

```
            do i=ii,min(ii+nb-1,n1)
              call fft(zwork(1,i-ii+1),n3)
            end do
! Step 6: Blocked transposition
            do k=1,n3
              do i=ii,min(ii+nb-1,n1)
                x(i,j,k)=zwork(k,i-ii+1)
              end do
            end do
          end do
        end do
      return
      end
```

8.3.4 Blocked Three-Dimensional FFT Algorithm

In the three-dimensional FFT algorithm, it is necessary to transpose the matrix three times, but the algorithm can be blocked in the same manner as the blocked six-step FFT and the blocked two-dimensional FFT. Listing 5 shows a blocked three-dimensional FFT based on the row–column algorithm.

Here, the parameters nb and np are the blocking parameter and padding parameter, respectively. The arrays ywork and zwork are the work arrays. In the blocked three-dimensional FFT algorithm, the array ywork of size $O(n_2)$ and the array zwork of size $O(n_3)$ are sufficient for the work arrays. In other words, the input data is overwritten with the output data, which is an in-place algorithm.

8.4 Parallel Three-Dimensional FFT Using Two-Dimensional Decomposition

As a typical array distribution method in parallel three-dimensional FFTs, thus far, only one dimension (for example, z-axis) among three dimensions (the x-, y-, and z-axes) is divided. In this case, the number of data in the z-axis must be greater than or equal to the number of MPI processes.

In the recent massively parallel cluster, the number of cores and the number of processors tend to increase in order to improve the performance. For example, Sunway TaihuLight, which is a massively parallel system, was ranked first in the TOP500 list of June 2017 [2], and the number of cores exceeds 10 million. In such a system, reducing the number of MPI processes by hybrid MPI and OpenMP parallel programming is effective in reducing the communication time. Even in that case, the maximum number of MPI processes is more than 10,000. Therefore, when one-dimensional decomposition is used in the z-axis, the number of data in the z-axis for such a system must be more than 10,000, and the problem size of the three-dimensional FFT is restricted.

As a method by which to solve this problem, a method of three-dimensional decomposition in the x-, y-, and z-axes has been proposed [7, 8]. In the three-dimensional decomposition, when performing FFTs in the x-, y-, and z-axes, it is necessary to exchange data by all-to-all communication each time. On the other hand, in the two-dimensional decomposition, since there is one undecomposed axis among the x-, y-, and z-axes, there is an advantage in that the number of all-to-all communications can be reduced [21]. Parallel three-dimensional FFT algorithms using two-dimensional decomposition have been proposed [1, 3, 14, 21].

In this section, we describe a parallel three-dimensional FFT algorithm using two-dimensional decomposition in the y- and z-axes. Using the parallel three-dimensional FFT algorithm, it is possible to obtain high scalability, even with a relatively small number of data.

8.4.1 Implementation of Parallel Three-Dimensional FFT Using Two-Dimensional Decomposition

Figure 8.5 shows the parallel three-dimensional FFT using one-dimensional block decomposition of initial data in the z-axis, which is a typical array distribution method in parallel three-dimensional FFT. In this subsection, a parallel three-dimensional FFT algorithm using two-dimensional decomposition in the y- and z-axes is described.

Assume that the number of data N is $N = N_1 \times N_2 \times N_3$, and the MPI process is mapped in two dimensions $P \times Q$. In a distributed memory parallel computer having $P \times Q$ MPI processes, the three-dimensional array $x(N_1, N_2, N_3)$ is distributed along the second dimension (N_2) and the third dimension (N_3). If N_2 is divisible by P and N_3 is also divisible by Q, $N_1 \times (N_2/P) \times (N_3/Q)$ pieces of data are distributed to each MPI process. Although somewhat complicated, we next introduce the notations $\hat{N}_r = N_r/P$ and $\hat{\hat{N}}_r = N_r/Q$. Moreover, $\hat{J}_r$ denotes that data along the index J_r is distributed to P MPI processes and $\hat{\hat{J}}_r$ denotes that data along the index J_r is distributed to Q MPI processes. Note that r means that the index is of dimension r. Thus, the distributed three-dimensional array can be expressed as $\hat{x}(N_1, \hat{N}_2, \hat{\hat{N}}_3)$. According to the block distribution, the local index $\hat{J}_r(l)$ in the l-th MPI process in the y-axis corresponds to the following global index, J_r:

$$J_r = l \times \hat{N}_r + \hat{J}_r(l), \quad 0 \leq l \leq P - 1, \quad 1 \leq r \leq 3. \tag{8.23}$$

Fig. 8.5 One-dimensional decomposition for a three-dimensional FFT

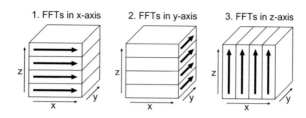

Moreover, the local index $\hat{\hat{J}}_r(m)$ in the m-th MPI process in the z-axis corresponds to the global index J_r as follows:

$$J_r = m \times \hat{\hat{N}}_r + \hat{\hat{J}}_r(m), \quad 0 \le m \le Q - 1, \quad 1 \le r \le 3. \quad (8.24)$$

Here, to show all-to-all communication, the notations $\tilde{N}_i \equiv N_i/P_i$ and $\tilde{\tilde{N}}_i \equiv N_i/Q_i$ are introduced. Here, N_i is decomposed into two-dimensional representations of $\tilde{N}_i$ and P_i and $\tilde{\tilde{N}}_i$ and Q_i. Note that P_i and Q_i are the same as P and Q, respectively, but this indicates that these indices are of dimension i.

Letting the initial data be $\hat{x}(N_1, \hat{N}_2, \hat{\hat{N}}_3)$, the parallel three-dimensional FFT using two-dimensional decomposition [21] is as follows:

Step 1: $(N_2/P) \cdot (N_3/Q)$ individual N_1-point multicolumn FFTs

$$\hat{x}_1(K_1, \hat{J}_2, \hat{\hat{J}}_3) = \sum_{J_1=0}^{N_1-1} \hat{x}(J_1, \hat{J}_2, \hat{\hat{J}}_3)\omega_{N_1}^{J_1 K_1}.$$

Step 2: Transposition

$$\hat{x}_2(\hat{J}_2, \hat{\hat{J}}_3, \tilde{K}_1, P_1) \equiv \hat{x}_2(\hat{J}_2, \hat{\hat{J}}_3, K_1)$$
$$= \hat{x}_1(K_1, \hat{J}_2, \hat{\hat{J}}_3).$$

Step 3: Q individual all-to-all communications across P processors in the y-axis

$$\hat{x}_3(\tilde{J}_2, \hat{\hat{J}}_3, \hat{K}_1, P_2) = \hat{x}_2(\hat{J}_2, \hat{\hat{J}}_3, \tilde{K}_1, P_1).$$

Step 4: Rearrangement

$$\hat{x}_4(J_2, \hat{\hat{J}}_3, \hat{K}_1) \equiv \hat{x}_4(\tilde{J}_2, P_2, \hat{\hat{J}}_3, \hat{K}_1) = \hat{x}_3(\tilde{J}_2, \hat{\hat{J}}_3, \hat{K}_1, P_2).$$

Step 5: $(N_3/Q) \cdot (N_1/P)$ individual N_2-point multicolumn FFTs

$$\hat{x}_5(K_2, \hat{\hat{J}}_3, \hat{K}_1) = \sum_{J_2=0}^{N_2-1} \hat{x}_4(J_2, \hat{\hat{J}}_3, \hat{K}_1)\omega_{N_2}^{J_2 K_2}.$$

Fig. 8.6 Two-dimensional decomposition for a three-dimensional FFT

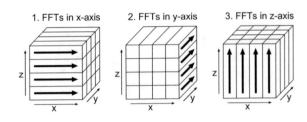

Step 6: Transposition

$$\hat{x}_6(\hat{\tilde{J}}_3, \; \hat{K}_1, \; \tilde{K}_2, \; Q_2) \equiv \hat{x}_6(\hat{\tilde{J}}_3, \; \hat{K}_1, \; K_2) = \hat{x}_5(K_2, \; \hat{\tilde{J}}_3, \; \hat{K}_1).$$

Step 7: P individual all-to-all communications across Q processors in the z-axis

$$\hat{x}_7(\tilde{\tilde{J}}_3, \; \hat{K}_1, \; \hat{\tilde{K}}_2, \; Q_3) = \hat{x}_6(\hat{\tilde{J}}_3, \; \hat{K}_1, \; \tilde{\tilde{K}}_2, \; Q_2).$$

Step 8: Rearrangement

$$\begin{aligned}
\hat{x}_8(J_3, \; \hat{K}_1, \; \hat{\tilde{K}}_2) &\equiv \hat{x}_8(\tilde{\tilde{J}}_3, \; Q_3, \; \hat{K}_1, \; \hat{\tilde{K}}_2) \\
&= \hat{x}_7(\tilde{\tilde{J}}_3, \; \hat{K}_1, \; \hat{\tilde{K}}_2, \; Q_3).
\end{aligned}$$

Step 9: $(N_1/P) \cdot (N_2/Q)$ individual N_3-point multicolumn FFTs

$$\hat{x}_9(K_3, \; \hat{K}_1, \; \hat{\tilde{K}}_2) = \sum_{J_3=0}^{N_3-1} \hat{x}_8(J_3, \; \hat{K}_1, \; \hat{\tilde{K}}_2)\omega_{N_3}^{J_3 K_3}.$$

Step 10: Transposition

$$\hat{y}(\hat{K}_1, \; \hat{\tilde{K}}_2, \; K_3) = \hat{x}_9(K_3, \; \hat{K}_1, \; \hat{\tilde{K}}_2).$$

Figure 8.6 shows the parallel three-dimensional FFT when the initial data is decomposed into two dimensions in the y- and z-axes. In this parallel three-dimensional FFT, when $N_1 = N_2 = N_3 = N^{1/3}$, $(N^{1/3}/P) \cdot (N^{1/3}/Q)$ individual $N^{1/3}$-point multicolumn FFTs are performed in Steps 1, 5, and 9. Note that while the input data $\hat{x}(J_1, \; \hat{J}_2, \; \hat{\tilde{J}}_3)$ is two-dimensionally decomposed in the y- and z-axes, the Fourier-transformed output data $\hat{y}(\hat{K}_1, \; \hat{\tilde{K}}_2, \; K_3)$ is two-dimensionally decomposed in the x- and y-axes. In this manner, by using different data distributions for input and output, it is sufficient to perform all-to-all communications twice in the y- and z-axes in Steps 3 and 7. When the same data distribution is used for input and output, it is necessary to perform all-to-all communications in the y- and z-axes once more.

8.4.2   Communication Time in One-Dimensional Decomposition and Two-Dimensional Decomposition

Let N be the total number of data, and let $P \times Q$ be the number of MPI processes. Moreover, let W be the communication bandwidth (byte/s) and let L be the communication latency (s). Hereinafter, the communication time in the cases of one-dimensional decomposition and two-dimensional decomposition will be examined. For simplicity, we assume that there is no communication contention in the all-to-all communications.

8.4.2.1   Communication Time in One-Dimensional Decomposition

In the case of one-dimensional decomposition, each MPI process will send $N/(PQ)^2$ pieces of double-precision complex data to $PQ - 1$ MPI process other than itself.

Therefore, the communication time $T_{1\text{dim}}$ in the one-dimensional decomposition is expressed as follows:

$$T_{1\text{dim}} = (PQ - 1)\left(L + \frac{16N}{(PQ)^2 \cdot W}\right)$$
$$\approx PQ \cdot L + \frac{16N}{PQ \cdot W}. \qquad (8.25)$$

8.4.2.2   Communication Time in Two-Dimensional Decomposition

In the case of two-dimensional decomposition, since Q pairs of all-to-all communications are performed among P MPI processes in the y-axis, each MPI process in the y-axis sends $N/(P^2 Q)$ double-precision complex data to $P - 1$ MPI processes in the y-axis. Moreover, since P pairs of all-to-all communications are performed among Q MPI processes in the z-axis, each MPI process in the z-axis sends $N/(PQ^2)$ double-precision complex data to $Q - 1$ MPI processes in the z-axis.

Therefore, the communication time $T_{2\text{dim}}$ in the two-dimensional decomposition is expressed as follows:

$$T_{2\text{dim}} = (P - 1)\left(L + \frac{16N}{P^2 Q \cdot W}\right) + (Q - 1)\left(L + \frac{16N}{PQ^2 \cdot W}\right)$$
$$\approx (P + Q) \cdot L + \frac{32N}{PQ \cdot W}. \qquad (8.26)$$

8.4.2.3 Comparison of Communication Time in One-Dimensional Decomposition and Two-Dimensional Decomposition

The communication times in the cases of one-dimensional decomposition and two-dimensional decomposition are expressed by Eqs. (8.25) and (8.26), respectively.

In comparing these two equations, the one-dimensional decomposition of Eq. (8.25) becomes approximately half of two-dimensional decomposition of Eq. (8.26) as the total communication amount. However, when the total number of MPI processes $P \times Q$ is large and the latency L is large, the communication time is shorter in the two-dimensional decomposition of Eq. (8.26). Here, a condition is obtained whereby the communication time $T_{2\text{dim}}$ represented by Eq. (8.26) is smaller than the communication time $T_{1\text{dim}}$ represented by Eq. (8.25).

From Eqs. (8.25) and (8.26),

$$(P + Q) \cdot L + \frac{32N}{PQ \cdot W} < PQ \cdot L + \frac{16N}{PQ \cdot W}. \tag{8.27}$$

We have

$$N < \frac{(LW \cdot PQ)(PQ - P - Q)}{16}. \tag{8.28}$$

For example, substituting $L = 10^{-5}$ (s), $W = 10^9$ (byte/s), and $P = Q = 64$ into Eq. (8.28), in the range of $N < 10^{10}$, the communication time of two-dimensional decomposition is found to be smaller than that of one-dimensional decomposition.

8.4.3 Performance Results

In the performance evaluation, we compare the performances of parallel three-dimensional FFT using two-dimensional decomposition and parallel three-dimensional FFT using one-dimensional decomposition. For the measurement, 32^3, 64^3, 128^3, and 256^3-point forward FFTs were performed 10 times and the average elapsed time was measured. The calculation of the FFT is performed with double-precision complex numbers and the trigonometric function table is prepared in advance. In multicolumn FFTs, the Stockham algorithm was used for in-cache FFT when each column FFT fits into the cache. The in-cache FFT is implemented by combinations of radices 2, 3, 4, 5, and 8, and the same program was used for one-dimensional decomposition and two-dimensional decomposition.

Moreover, in order to reduce the number of all-to-all communications, different data distribution formats are used for the input and output of the three-dimensional FFT. In the one-dimensional decomposition, the input is decomposed in the z-axis and the output is decomposed in the x-axis. In the two-dimensional decomposition, the input is decomposed in the y- and z-axes and the output is decomposed in the x- and y-axes.

Fig. 8.7 Performance of parallel three-dimensional FFTs using two-dimensional decomposition

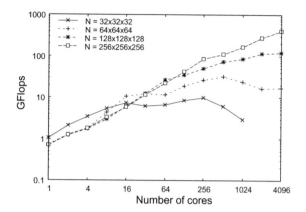

An Appro Xtreme-X3 (648 nodes, 32 GB per node, 147.2 GFlops per node, total main memory: 20 TB, communication bandwidth: 8 GB/s per node, and peak performance: 95.4 TFlops) was used as a multicore massively parallel cluster. Each computation node of the Xtreme-X3 has a four-socket quad-core AMD Opteron 8356 (Barcelona, 2.3 GHz) in a 16-core shared-memory configuration with a performance of 147.2 GFlops. All of the nodes in the system are connected through a full-bisectional fat tree network with DDR InfiniBand.

MVAPICH 1.2.0 [12] was used as a communication library. In this performance evaluation, one MPI process per core was used and the flat MPI programming model was used because this evaluation focuses on scalability when the number of MPI processes is large. The Intel Fortran Compiler 10.1 was used with the compiler option "ifort -O3 -xO".

Figure 8.7 shows the performance of parallel three-dimensional FFTs using two-dimensional decomposition. Here, the GFlops value of $N = 2^m$-point FFT is calculated from $5N \log_2 N$. As shown in Fig. 8.7, good scalability is not obtained using 32^3-point FFT. This is because the problem size is small (1 MB), which is why all-to-all communication dominates most of the total execution time. On the other hand, the performance is improved up to 4,096 cores for 256^3-point FFT. The performance for 4,096 cores was approximately 401.3 GFlops.

The performances of one-dimensional decomposition and two-dimensional decomposition in 256^3-point parallel three-dimensional FFT are shown in Fig. 8.8. As shown in Fig. 8.8, in the case of 64 or fewer cores, the one-dimensional decomposition with a small communication amount has higher performance than the two-dimensional decomposition. On the other hand, for 128 or more cores, the two-dimensional decomposition, which can reduce the communication time, has higher performance than the one-dimensional decomposition.

In the case of one-dimensional decomposition, 256^3-point FFT can be performed in 256 MPI processes or fewer, but 256^3-point FFT can be performed in up to 65,536 MPI processes when two-dimensional decomposition is used.

Fig. 8.8 Performance of parallel three-dimensional 256^3-point FFTs

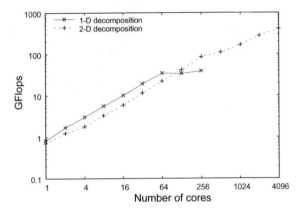

Table 8.2 Execution and communication time of parallel 256^3-point FFTs using two-dimensional decomposition

Number of cores	Execution time (s)	Communication time (s)	% of communication
1	2.84448	0.33756	11.87
2	1.60708	0.25705	15.99
4	1.11583	0.26579	23.82
8	0.61128	0.16514	27.02
16	0.34474	0.11832	34.32
32	0.17320	0.06372	36.79
64	0.09039	0.03737	41.34
128	0.04823	0.02176	45.11
256	0.02346	0.01320	56.28
512	0.01793	0.01461	81.48
1024	0.01199	0.01079	89.98
2048	0.00736	0.00652	88.64
4096	0.00502	0.00482	96.07

Table 8.2 shows the ratio of the communication time to the execution time for 256^3-point parallel three-dimensional FFT using two-dimensional decomposition. As shown in Table 8.2, for 4,096 cores, more than 96% of the total time is taken up by communication. The reason for this is considered to be that latency is dominant in the communication time because the amount of communications data sent by each MPI process at one time in all-to-all communication is only 1 KB.

In order to further reduce the communication time, it is necessary to reduce the latency by using a lower level communication function without using the `MPI_Alltoall` function for all-to-all communication, as in reference [7].

8.5 Parallel One-Dimensional FFT in a GPU Cluster

In recent years, the high performance and memory bandwidth of the graphics processing unit (GPU) have attracted attention, and attempts have been made to apply the GPU to various HPC applications. Moreover, GPU clusters, which connect a large number of computation nodes equipped with GPUs, are also in widespread use, and Titan, which is a GPU cluster equipped with NVIDIA Tesla K20X GPUs, was ranked fourth in the TOP500 list of June 2017 [2].

In performance tuning of such a GPU cluster, it is difficult to manually tune the optimum performance parameters, which depend on the architecture of the GPU, the network connecting the nodes, the problem size, and so on, each time.

FFTW [10] and SPIRAL [16] have been proposed as FFT libraries to which automatic tuning is applied in a distributed memory parallel computer, but FFT libraries to which automatic tuning is applied in GPU clusters have not yet been proposed.

In this section, we describe the results of a performance evaluation for applying automatic tuning to a parallel one-dimensional FFT in a GPU cluster.

Listing 6 CUDA Fortran program of parallel one-dimensional FFT [22]

```
        complex(8)  ::  A(N/P),B(N/P)
        complex(8),device  ::  A_d(N/P),B_d(N/P)
        integer  ::  plan1,plan2
! Decompose N into N1 and N2
        call GETN1N2(N,N1,N2)
! Create 1-D FFT plans
        istat=cufftPlan1D(plan1,N1,CUFFT_Z2Z,N2/P)
        istat=cufftPlan1D(plan2,N2,CUFFT_Z2Z,N1/P)
! Copy data from host memory to device memory
        A_d=A
! Step 1: Rearrange (N1/P)*P*(N2/P) to (N1/P)*(N2/P)*P
        call REARRANGE(A_d,B_d,N1/P,P,N2/P)
! Step 2: All-to-all communication
between GPU and GPU
        call MPI_ALLTOALL(B_d,N/P,MPI_DOUBLE_COMPLEX,A_d,N/P,
     &                    MPI_DOUBLE_COMPLEX,MPI_COMM_WORLD,ierr)
! Step 3: Transpose (N1/P)*N2 to N2*(N1/P)
        call TRANSPOSE(A_d,B_d,N1/P,N2)
! Step 4: (N1/P) individual N2-point multicolumn
FFTs
        istat=cufftExecZ2Z(plan2,B_d,B_d,CUFFT_FORWARD)
! Step 5: Rearrange (N2/P)*P*(N1/P) to (N2/P)*(N1/P)*P and
!         twiddle factor multiplication
        call REARRANGE_TWIDDLE(B_d,A_d,N2/P,P,N1/P)
! Step 6: All-to-all communication
between GPU and GPU
        call MPI_ALLTOALL(A_d,N/P,MPI_DOUBLE_COMPLEX,B_d,N/P,
     &                    MPI_DOUBLE_COMPLEX,MPI_COMM_WORLD,ierr)
! Step 7: Transpose (N2/P)*N1 to N1*(N2/P)
        call TRANSPOSE(B_d,A_d,N2/P,N1)
! Step 8: (N2/P) individual N1-point multicolumn FFTs
        istat=cufftExecZ2Z(plan1,A_d,A_d,CUFFT_FORWARD)
! Step 9: Rearrange (N1/P)*P*(N2/P) to (N1/P)*(N2/P)*P
        call REARRANGE(A_d,B_d,N1/P,P,N2/P)
! Step 10: All-to-all communication between GPU
and GPU
        call MPI_ALLTOALL(B_d,N/P,MPI_DOUBLE_COMPLEX,A_d,N/P,
     &                    MPI_DOUBLE_COMPLEX,MPI_COMM_WORLD,ierr)
```

```
! Step 11: Transpose (N1/P)*N2 to N2*(N1/P)
      call TRANSPOSE(A_d,B_d,N1/P,N2)
! Copy data from device memory to host memory
      B=B_d
! Destroy the 1-D FFT plans
      istat=cufftDestroy(plan1)
      istat=cufftDestroy(plan2)
```

8.5.1 Implementation of Parallel One-Dimensional FFT in a GPU Cluster

Implementation of parallel one-dimensional FFT in a GPU cluster has been proposed [22], but all-to-all communication takes place three times when performing parallel one-dimensional FFT, so most of the computation time is dominated by all-to-all communication. Furthermore, since the theoretical bandwidth of the PCI Express bus, which is the interface connecting the CPU and the GPU, is 8 GB/s per direction in the PCI Express Gen2 x 16 lanes. Thus, it is also important to reduce the amount of data transfer between the CPU and the GPU.

When transferring the memory on the GPU by MPI, basically, the following procedure is basically necessary:

(1) Copy data from the device memory on the GPU to the host memory on the CPU.
(2) Transfer data using the MPI communication function.
(3) Copy data from the host memory on the CPU to the device memory on the GPU.

In this case, there is a problem in that MPI communication is not performed while data is transferred between the CPU and the GPU. Therefore, this problem was solved using the MVAPICH2-GPU [25], which is an MPI library that can pipeline data transfer between a CPU and a GPU and MPI communication between nodes and overlap. Furthermore, we devised a process such that a matrix transposition appearing in FFT processing is performed on the GPU.

Listing 6 shows a parallel one-dimensional FFT program using CUDA Fortran [15].

8.5.2 Automatic Tuning

When performance tuning a parallel one-dimensional FFT, the following are the three main performance parameters [20]:

(1) All-to-all communication methods,
(2) Radices, and
(3) Block sizes.

By searching for these performance parameters, it is possible to further improve the performance of the parallel one-dimensional FFT. Note that (1) is a parameter related to MPI inter-process communication, and (2) and (3) are parameters related to performance within the MPI process. Therefore, the tuning of (1) and the tuning of (2) and (3) can be performed independently.

8.5.2.1   All-to-All Communication Methods

Research to automatically tune collective communication of MPI has been conducted [9]. Moreover, in multicore clusters connected by InfiniBand, a method of improving the performance by dividing all-to-all communication into two steps, intra-node exchange and inter-node exchange, has also been proposed [11]. A generalized two-step all-to-all communication algorithm [20], in which P MPI processes can be decomposed into $P = P_x \times P_y$, is as follows:

Let N be the total number of elements of the array of all MPI processes.

1. In each MPI process, copy the array subscript order so that it changes from $(N/P^2, P_x, P_y)$ to $(N/P^2, P_y, P_x)$. Next, P_x pairs of all-to-all communications among P_y MPI processes are performed.
2. In each MPI process, copy the array subscript order so that it changes from $(N/P^2, P_y, P_x)$ to $(N/P^2, P_x, P_y)$. Next, P_y pairs of all-to-all communications among P_x MPI processes are performed.

In this two-step all-to-all communication algorithm, since all-to-all communication among the nodes is performed twice, the total communication amount is doubled compared to the case in which all MPI processes are communicated among P MPI processes. However, since the start-up time of all-to-all communication is proportional to the number of MPI processes P, when N is relatively small, and the number of MPI processes P is large, compared with a simple all-to-all communication algorithm, the two-step all-to-all algorithm may be advantageous.

Therefore, by searching all combinations of P_x and P_y such that $P = P_x \times P_y$, optimum combinations of P_x and P_y can be examined. When the number of MPI processes P is a power of two, even if all combinations of P_x and P_y are tried, the search space is at most $\log_2 P$. Figure 8.9 shows the automatic tuning method for all-to-all communication.

Table 8.3 shows the results of applying automatic tuning to all-to-all communication using 32 nodes and 128 MPI processes in the HA-PACS base cluster. Table 8.3 shows that the two-step all-to-all communication is selected with message sizes in the range of 1–2 KB and 32–256 KB. Since 16 cores are equipped in one node of the HA-PACS base cluster used for evaluation, when $P_x = 1, 2, 4$, closed communication within the node at the first step is performed among P_x MPI processes. When the message sizes are 1, 32, and 64 KB, $P_x = 8$ and $P_y = 16$ are selected, which shows that two-step all-to-all communication is performed among the nodes.

Fig. 8.9 Automatic tuning
of all-to-all communication

```
min_time = DBL_MAX;
for (i = 0; i <= log₂(P); i++) {
  Pₓ = 2ⁱ;
  Pᵧ = P/Pₓ;
  MPI_Barrier(MPI_COMM_WORLD);
  start = MPI_Wtime();
  for (count = 0; count < ITER_NUM; count++) {
    if (Pₓ == 1 || Pᵧ == 1)
      MPI_Alltoall(sendbuf, ..., recvbuf, ...);
    else
      Two-Step-Alltoall(sendbuf, ..., recvbuf, ..., Pₓ, Pᵧ, ...);
  }
  MPI_Barrier(MPI_COMM_WORLD);
  end = MPI_Wtime();
  if (end − start < min_time) {
    min_time = end − start;
    Qₓ = Pₓ;
    Qᵧ = Pᵧ;
  }
}
Pₓ = Qₓ;
Pᵧ = Qᵧ;
```

Table 8.3 Performance of all-to-all communication (HA-PACS base cluster, 32 nodes, 128 MPI processes)

Message size (bytes)	MPI_Alltoall	Automatically tuned all-to-all		
	Time (s)	P_x	P_y	Time (s)
16	0.00070	128	1	0.00060
32	0.00068	128	1	0.00064
64	0.00071	128	1	0.00070
128	0.00079	128	1	0.00074
256	0.00097	128	1	0.00099
512	0.00172	128	1	0.00161
1024	0.00234	8	16	0.00206
2048	0.00479	4	32	0.00343
4096	0.00319	128	1	0.00324
8192	0.00669	128	1	0.00482
16384	0.00939	128	1	0.00751
32768	0.02326	8	16	0.02322
65536	0.04321	8	16	0.04009
131072	0.08963	4	32	0.07588
262144	0.13694	2	64	0.12529
524288	0.24869	1	128	0.21937
1048576	0.43614	128	1	0.39528
2097152	0.79572	1	128	0.78185
4194304	1.59248	128	1	1.51072

8.5.2.2   Radices

In the six-step FFT algorithm, the number of data N is decomposed into $N = N_1 \times N_2$ in order to calculate N_1 individual N_2-point FFTs and N_2 individual N_1-point FFTs, respectively. Let us call the radices N_1 and N_2. The values of N_1 and N_2 can be chosen arbitrarily as long as $N = N_1 \times N_2$ is satisfied. However, in GPU clusters, when P is the number of MPI processes, it is necessary to satisfy N_1, $N_2 \geq P$.

Usually, N_1 and N_2 are chosen to be $N_1 \approx N_2 \approx \sqrt{N}$, but we can choose N_1 and N_2 so that the performance is the highest. If N and P are powers of two, N_1 is varied with P, $2P$, ..., $\sqrt{N}$, and then $N_2 = N/N_1$. In this case, even if all combinations of N_1 and N_2 are tried, the search space is $\log_2(\sqrt{N}/P)$.

8.5.2.3   Block Sizes

In the six-step FFT algorithm, it is necessary to transpose the matrix, but it is known that transposition of this matrix can be performed efficiently by performing cache blocking. At this time, the optimum block size NB depends on the problem size, the cache size, and other factors.

In this implementation, the block size NB is limited to powers of two and is varied as 16, 32, 64, 128, 256, 512, and 1024. Although the optimum block size NB is not necessarily a power of two, the automatic tuning method is considered to be effective even when the block size NB is other than a power of two. An automatic tuning method of parallel one-dimensional FFT is shown in Fig. 8.10.

Fig. 8.10 Automatic tuning of parallel one-dimensional FFT

```
min_time = DBL_MAX;
for (j = 4; j <= 10; j++) {
  NB = 2^j;
  for (i = log_2(P); i <= log_2(√N); i++) {
    N_1 = 2^i;
    N_2 = N/N_1;
    MPI_Barrier(MPI_COMM_WORLD);
    start = MPI_Wtime();
    for (count = 0; count < ITER_NUM; count++) {
      Parallel_1D_FFT(...);
    }
    MPI_Barrier(MPI_COMM_WORLD);
    end = MPI_Wtime();
    if (end − start < min_time) {
      min_time = end − start;
      M_1 = N_1;
      M_2 = N_2;
      NB2 = NB;
    }
  }
}
N_1 = M_1;
N_2 = M_2;
NB = NB2;
```

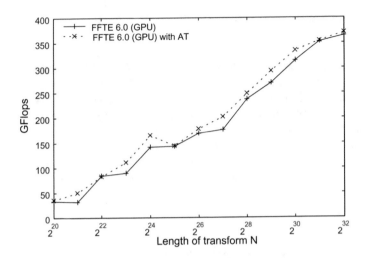

Fig. 8.11 Performance of parallel one-dimensional FFT (HA-PACS base cluster, 32 nodes, 128 MPI processes)

8.5.3 Performance Results

In the performance evaluation, performance comparison between FFTE (version 6.0) [1], which is an FFT library supporting GPU clusters, and the automatic tuning method was applied to FFTE. We changed m of $N = 2^m$ and then executed forward FFTs ten times and measured the average elapsed time. The calculation of the FFT was with double-precision complex numbers and NVIDIA CUFFT [13] was used as an FFT routine for a single GPU.

An HA-PACS base cluster was used as a GPU cluster. The HA-PACS base cluster is a GPU cluster with Appro Xtreme-X consisting of 268 nodes, 4288 cores, and 1072 GPUs. Each node is equipped with two sockets of an Intel Xeon E5-2670 (Sandy Bridge-EP 2.6 GHz), four NVIDIA M2090 as a GPU, and nodes connected by a fat tree network with two rails of InfiniBand QDR. In the experiment, we used from 1 to 32 nodes. The parallel one-dimensional FFT programs were run on from 1 to 128 MPI processes, i.e., each node has four MPI processes.

MVAPICH2 2.0b was used as a communication library. The compiler used PGI Fortran compiler 14.2 and the option "pgf90 -fast -Mcuda=cc2x, cuda5.5".

Figure 8.11 shows the performances of FFTE 6.0 and FFTE 6.0 with automatic tuning. Here, the unit of the execution time is seconds and the GFlops value of $N = 2^m$-point FFT is calculated from $5N \log_2 N$. Figure 8.11 shows that the performance is improved by applying automatic tuning to FFTE 6.0.

---

[1]http://www.ffte.jp/

8.6 Summary

In this chapter, optimization of FFT in a large-scale system was described. In the blocked FFT algorithm, as a result of reducing the cache miss by increasing the reuse ratio of data in the cache memory, the number of accesses to the main memory can also be reduced. The blocked FFT algorithms are more effective when the difference between the processing speed of the processor and the access speed of the main memory is large.

It is not easy to improve the efficiency of the FFT as the size of parallel supercomputers increases. Two-dimensional decomposition and three-dimensional decomposition are considered to be essential.

Exercises

1. Calculate Eq. (8.1) for $x(j) = j$, $0 \leq j \leq 3$ and find the number of real multiplications and the number of real additions.
2. Calculate the same calculation by reducing Eqs. (8.9) and (8.10) to the two-point DFT. Then, calculate the number of real multiplications and the number of real additions, and compare the results.
3. Develop programs to calculate DFT and FFT in arbitrary programming languages. Then, compare the computation time required for a 65,536-point complex Fourier transform.

References

1. 2DECOMP & FFT—Library for 2D Pencil Decomposition and Distributed FFTs. http://www.2decomp.org/
2. TOP500 Supercomputer Sites. http://www.top500.org/
3. O. Ayala, L.P. Wang, Parallel implementation and scalability analysis of 3D fast Fourier transform using 2D domain decomposition. Parallel Comput. **39**, 58–77 (2013)
4. D.H. Bailey, FFTs in external or hierarchical memory. J. Supercomput. **4**, 23–35 (1990)
5. W.T. Cochran, J.W. Cooley, D.L. Favin, H.D. Helms, R.A. Kaenel, W.W. Lang, G.C. Maling, D.E. Nelson, C.M. Rader, P.D. Welch, What is the fast Fourier transform? IEEE Trans. Audio Electroacoust. **15**, 45–55 (1967)
6. J.W. Cooley, J.W. Tukey, An algorithm for the machine calculation of complex Fourier series. Math. Comput. **19**, 297–301 (1965)
7. M. Eleftheriou, B.G. Fitch, A. Rayshubskiy, T.J.C. Ward, R.S. Germain, Scalable framework for 3D FFTs on the Blue Gene/L supercomputer: implementation and early performance measurements. IBM J. Res. Dev. **49**, 457–464 (2005)
8. B. Fang, Y. Deng, G. Martyna, Performance of the 3D FFT on the 6D network torus QCDOC parallel supercomputer. Comput. Phys. Commun. **176**, 531–538 (2007)
9. A. Faraj, X. Yuan, Automatic generation and tuning of MPI collective communication routines, in *Proceedings of 19th ACM International Conference on Supercomputing (ICS'05)* (2005), pp. 393–402
10. M. Frigo, S.G. Johnson, The design and implementation of FFTW3. Proc. IEEE **93**, 216–231 (2005)

11. R. Kumar, A. Mamidala, D.K. Panda, Scaling all-to-all collective on multi-core systems, in *Proceedings of 2008 IEEE International Parallel and Distributed Processing Symposium (IPDPS 2008)* (2008)
12. MVAPICH: MPI over InfiniBand and iWARP. http://mvapich.cse.ohio-state.edu/
13. NVIDIA Corporation, CUFFT Library User's Guide (2017). http://docs.nvidia.com/cuda/pdf/CUFFT_Library.pdf
14. D. Pekurovsky, P3DFFT: a framework for parallel computations of Fourier transforms in three dimensions. SIAM J. Sci. Comput. **34**, C192–C209 (2012)
15. The Portland Group, CUDA Fortran Programming Guide and Reference (2017). http://www.pgroup.com/doc/pgicudaforug.pdf
16. M. Püschel, J.M.F. Moura, J.R. Johnson, D. Padua, M.M. Veloso, B.W. Singer, J. Xiong, F. Franchetti, A. Gačić, Y. Voronenko, K. Chen, R.W. Johnson, N. Rizzolo, SPIRAL: code generation for DSP transforms. Proc. IEEE **93**, 232–275 (2005)
17. G.E. Rivard, Direct fast Fourier transform of bivariate functions. IEEE Trans. Acoust. Speech Signal Process. **ASSP-25**, 250–252 (1977)
18. R.C. Singleton, An algorithm for computing the mixed radix fast Fourier transform. IEEE Trans. Audio Electroacoust. **17**, 93–103 (1969)
19. D. Takahashi, A blocking algorithm for parallel 1-D FFT on shared-memory parallel computers, in *Proceedings of 6th International Conference on Applied Parallel Computing (PARA 2002)*. Lecture Notes in Computer Science, vol. 2367 (Springer, 2002), pp. 380–389
20. D. Takahashi, Automatic tuning for parallel FFTs, in *Software Automatic Tuning: From Concepts to State-of-the-Art Results*, ed. by K. Naono, K. Teranishi, J. Cavazos, R. Suda (Springer, 2010), pp. 49–67
21. D. Takahashi, An implementation of parallel 3-D FFT with 2-D decomposition on a massively parallel cluster of multi-core processors, in *Proceedings of 8th International Conference on Parallel Processing and Applied Mathematics (PPAM 2009), Part I, Workshop on Memory Issues on Multi- and Manycore Platforms*. Lecture Notes in Computer Science, vol. 6067 (Springer, 2010), pp. 606–614
22. D. Takahashi, Implementation of parallel 1-D FFT on GPU clusters, in *Proceedings of 2013 IEEE 16th International Conference on Computational Science and Engineering (CSE 2013)* (2013), pp. 174–180
23. C. Temperton, Self-sorting mixed-radix fast Fourier transforms. J. Comput. Phys. **52**, 1–23 (1983)
24. C. Van Loan, *Computational Frameworks for the Fast Fourier Transform* (SIAM Press, Philadelphia, PA, 1992)
25. H. Wang, S. Potluri, M. Luo, A.K. Singh, S. Sur, D.K. Panda, MVAPICH2-GPU: optimized GPU to GPU communication for infiniband clusters. Comput. Sci.—Res. Dev. **26**, 257–266 (2011)

Chapter 9
Optimization and Related Topics

Hiroshi Watanabe

Abstract In this chapter, we cover some topics related to optimization and debugging, including debugging techniques, usage of profilers, and application of version control systems.

9.1 Introduction

Before starting this chapter, let us consider why we want to optimize and/or parallelize codes. Both optimizing and parallelizing are a method to shorten the execution time keeping its behavior. Suppose there is a program for which the total execution time is expected to be 10 days, and it allows speeding up by 10%, then we can spend only 1 day for the optimization, otherwise, the time spent on the optimization is longer than the time we gain from the optimization. Usually, improving the technique of debugging, instead of optimization, contributes more to shorten the time of the study. Most of the development time is usually spent on debugging. Additionally, it is hopelessly difficult to debug a parallel code. Conversely, a programmer who writes program almost without bugs is much more productive compared with a programmer who writes buggy codes. In this chapter, we introduce several debugging techniques and related topics for the purpose of shortening the total time of the study.

9.2 Bug-Free Development

9.2.1 Basics of Debugging

The most time-consuming part of the development of the program is debugging. While it would be wonderful if you do not introduce any bugs, bug-free programming is almost impossible. Since bugs are unavoidable, we have to find and fix them as

H. Watanabe (✉)
Faculty of Science and Technology, Dept. Applied Physics and Physico-Informatics,
Keio University, Tokyo, Japan
e-mail: hwatanabe@appi.keio.ac.jp

© Springer Nature Singapore Pte Ltd. 2019
M. Geshi (ed.), *The Art of High Performance Computing for Computational
Science, Vol. 1*, https://doi.org/10.1007/978-981-13-6194-4_9

soon as possible in order to achieve rapid developments. There are two kinds of bugs, an easy-to-find bug, and a hidden bug. An easy-to-find bug is a bug which is found immediately when it is introduced. A hidden bug is a bug which is in hiding for a while and causes a problem later. It is usually more difficult, and therefore, more important to find and fix a hidden bug. But we first consider how to find and fix an easy-to-find bug. Since debugging is the most important part of programming, there is a long history to avoid, find, and fix bugs. Many techniques or tools have been developed such as test-driven development, agile software development, extreme programming, and so forth. If you have wielded such sophisticated tools, then you can skip this section. Here, we introduce some basics of debugging, especially focusing on unit testing and sort-diff debugging.

Unit testing is one of software development methods in which the behavior of small parts of the program are individually and independently tested. The testing process is performed, for example, as follows: (i) pick a part of the program which you want to check the behavior, (ii) setup environment so that the part can be compiled and executed without other routines, (iii) and prepare inputs and compare the results with the expected results. Especially, we have to confirm that the behavior does not change between before and after optimizations.

The sort-diff debugging includes the following procedures: (i) prepare two programs, for example, before and after optimizations, (ii) run two programs and sort both results, (iii) and compare the sorted results.

The sort-diff debugging is a kind of print debugging. Print debugging is a method to trace the status of the program by inserting printf functions in the source codes.[1] While print debugging is classical and looks an old-fashioned technique, this is a basic technique of debugging. Therefore, it would be better to see how it works before pursuing more sophisticated techniques or tools. The sort-diff debugging is usually used with unit testing. In the following, we introduce several specific examples of sort-diff debugging taking optimization in molecular dynamics (MD) as an example.

9.2.2 Sort + diff Debugging 1: Pair List Construction

Consider particle models with a finite cutoff length, such as MD or Smoothed Particle Hydrodynamics (SPH). To perform the time evolution of a system, we first have to find particles/atoms pairs such that the distance between them is less than the cutoff length. The list which contains the pairs of interacting particles/atoms are called a pair list, which is also called a Verlet list. The simplest way to construct a pair list is to check the distances of all pairs. This algorithm is shown in Algorithm 1.

---

[1]Therefore, it is also called printf debugging.

Algorithm 1 Construction of pair list with $O(N^2)$

1: **for** $i \leftarrow 1, N-1$ **do**
2:    **for** $j \leftarrow i+1, N$ **do**
3:        $r^2 \leftarrow (\mathbf{q}_i - \mathbf{q}_j)^2$
4:        **if** $r^2 < r_c^2$ **then**
5:            Append (i, j) to a pair list
6:        **end if**
7:    **end for**
8: **end for**

Fig. 9.1 Construct
information of grid. The case
for the two-dimensional
system is shown for
visibility. A system is
divided into cells with the
linear length l. A serial
number is assigned to each
cell. The atoms do not
interact with the atoms in the
next-nearest cells when the
cell size l is longer than the
cutoff length r_c. For
instance, the atoms in cell 4
will not interact with the
atoms in cell 3, 6, and 9

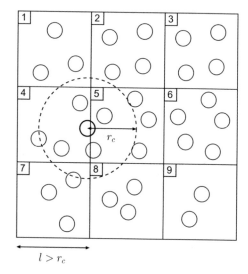

$l > r_c$

You can find that Algorithm 1 is simple, easy to implement. However, this algorithm is unpractical because the complexity of this algorithm is $O(N^2)$ for a number of atoms N. In order to reduce the complexity to $O(N)$, we adopt a grid search method, which is also called the linked-list method [1]. The algorithm includes the following steps: (i) divide a system into small cells. (ii) register the indices of atoms on cells to which they belong, and (iii) search for interacting atoms pairs using the registered information. See Fig. 9.1 for a schematic representation of the grid search. The complexities of constructing cell information and searching for interacting atoms using the information are both $O(N)$. However, the implementation of the grid search algorithm requires much more lines compared with the naive implementation of $O(N^2)$.

Generally, the possibility of inserting bugs increases as the total number of lines of source codes. The naive algorithm of $O(N^2)$ can be implemented in twenty lines, and therefore, the possibility of containing bug is expected to be very small. Therefore, we can use the $O(N^2)$ code as the reference to make the $O(N)$ code.

List 9.1 Check the consistency of the two methods

```
$ g++ on2code.cpp -o on2code
$ g++ on1code.cpp -o on1code
$ ./on2code | sort >on2.dat
$ ./on1code | sort >on1.dat
$ diff on1.dat on2.dat
```

The developing process is as follows:

1. Prepare the program which only contains the routines making the initial position of atoms and constructing pair lists (Unit testing).
2. Put the identical initial condition to $O(N^2)$ and $O(N)$ routines. Then two pair lists will be generated.
3. Confirm that the two pair lists are identical. Since the order of pair list may vary, compare them after sorting.

We use a sort command for sorting and use diff command to compare two results, and therefore, this method is called sort-diff debugging. The point is that the identical pair lists should be generated from the identical initial condition regardless of the construction algorithm. Suppose on2code.cpp and on1code.cpp are the programs of $O(N^2)$ and $O(N)$, respectively. The testing procedure is shown in List 9.1. If the two results are identical, then nothing is printed by diff. But it may print some differences at the beginning of development. The detailed information, such as which pair is overlooked, will be helpful for debugging the $O(N)$ code.

9.2.3 Sort + diff Debugging 2: Exchanging Information of Atoms

A supercomputer consists of many nodes, a node consists of several CPUs, and a CPU has many CPU-cores. Therefore, parallelization is unavoidable to utilize the computational power of supercomputers to the full. For MD simulations with short-range interaction, the domain decomposition is usually adopted for parallelization, i.e., the system is divided into several domains and each domain is assigned to a process. In order to calculate the force between atoms which are assigned to different processes, communication is necessary. Consider a simple decomposition of the three-dimensional system. Because there are 26 neighboring domains, a naive implementation includes 26 communications. The naive implementation is shown in Fig. 9.2a. Note that, the case of the two-dimensional system is shown for the visibility.

The number of communications can be reduced by transferring information of atoms received from other processes. See Fig. 9.2b. First, domain A sends information of atoms on the edge to domain B. Then domain B sends the atoms both on the edge of domain B itself and the atoms received from domain A. After the above two

List 9.2   Check the consistency of the two methods

```
$ cd test1; ./a.out; cd ..
$ cd test2; ./b.out; cd ..
$ diff <(sort test1/proc000.dat) <(sort test2/proc000.dat)
$ diff <(sort test1/proc001.dat) <(sort test2/proc001.dat) ...
```

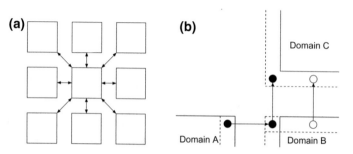

Fig. 9.2   a Simple communication. **b** Transfer method with fewer number of communications

steps, domain C obtains the atoms from domain A. The number of communications is reduced to be 6 from 26 for the three-dimensional system. The point is that the program including 26 communications is easier to be implemented compared with the one including 6 times of communication. Therefore, we use the naive implementation as the reference for making a program with the transfer. The development procedure is as follows:

1. Prepare the program which only contains the routines making initial positions and communication (Unit testing).
2. After communications, make a pair list on each process.
3. Print the pair lists to files. For example, the pair list in process 12 is exported to a file `proc012.dat`, and so forth.
4. Compare the results between the naive method and the method with transfer information (Sort + diff debugging).

Suppose `a.out` is an execution binary of the naive method and `b.out` is one adopting the information transfer. The test procedure is shown in List 9.2. You may think it is bothersome to implement two kinds of algorithms, one including 26 times of communication and the other including 6 times. However, it would be better to debug the parallel code seriously in the early stages of the development, since the debugging of the parallel code is hopelessly difficult.

9.2.4   Sort + diff Debugging 3: Parallel Pair List Construction

To perform a parallel MD simulation with the domain decomposition strategy, we have to construct pair lists in each domain, i.e., we have to parallelize pair list constructions. Obviously, the identical pair list should be generated from the identical

List 9.3 Check consistency between serial and parallel codes

```
$ ./serial | sort > serial.dat
$ ./parallel
$ cat proc*.dat | sort > parallel.dat
$ diff serial.dat parallel.dat
```

input, regardless of whether the code is serial or parallel. We utilize this fact to make the parallel pair list construction codes as follows:

1. Prepare the program which only contains the routines of making initial positions, communication, and pair list constructions (Unit testing).
2. Make a serial program which outputs a pair list.
3. Make a parallel program that outputs a pair list when atoms with small indices are assigned to a process. For example, a process 12 will generates proc012.dat.
4. Sort the results of the parallel code after concatenating and taking diff with the sorted result of the serial code (Sort + diff debugging).

Suppose serial and parallel are execution binaries of serial and parallel codes, respectively. The test procedure is shown in List 9.3.

At this point, we are sure that the serial pair list construction codes and communication codes are reliable because we have already debugged them. The most important point of the unit testing is isolation, i.e., we have to examine only one behavior at a time. After debugging, we can consider the routine as a comfort zone, and we can exclude it from routines in which a bug is hiding. After all, such a step-by-step development is the shortest path to develop a bug-free program.

9.2.5 Summary

The most important part of programming is debugging, and the most important part of debugging is to identify the position of the bug. For that, it is important to follow the rule that "test single behavior at a time," which is the main policy of the unit testing. If you want to check the pair list construction, then you should check it only. Do not perform time evolution because time evolution includes various behaviors such as force calculation, energy observation, updating position and momenta, and so forth. You should debug with microscopic information, not macroscopic information. Microscopic information includes, for example, the position of atoms, all indices of interacting atoms, and so forth. The typical macroscopic value is the total energy of the system. The total energy may be apparently conserved even if the program contains some bugs. If you perform debugging perfunctorily, then you shall pay for it later. In the present chapter, we introduce sort + diff debugging, which is one of the rather classical techniques for debugging. More sophisticated techniques have been developing, such as the test-driven development (TDD) [2], the extreme programming (XP) [3], the agile software development [4], and so forth. Refer the references to your next step.

9.3 Hidden Bugs

As described above, there are two kinds of bugs. One is the easy-to-find which can be found immediately and the other is hidden which appears later. It is usually difficult to find and fix the hidden bugs and there is no silver bullet for them. In this section, we will see a typical example of the hidden bug. We also introduce a method of debugging with the help of a version control system.

9.3.1 Example of Hidden Bugs

Here we introduce an example of a hidden bug which the author actually experienced. Suppose a code exited abnormally with a segmentation fault. First, we have to identify at which point the program terminates. One of the simplest ways to identify the position of the problem is a binary search using printf.[2] Put three printf statements that prints 1, 2 and 3 as follows:

```
printf ("1");
...
printf ("2");
...
printf ("3");
```

Note that, the first and last printf should be inserted so that the position of the problem surely exists between them. If the program stopped with printing only "1", then the problem exists between 1 and 2. If the program stopped with "12", then the problem exists between 2 and 3. Repeating the above procedures, we can identify the position of the problem quickly.

Then, suppose the problem is found to be accessing an array out of bounds.

```
const int N =10;
double data[N];
....
double func(int index){
   return data[index]; // <- index was 10 here.
}
```

---

[2]Of course you can use a debugger such as gdb if you want.

In the above example, the value of index should be from 0 to 9 but it was 10. The real problem occurs where this invalid value is set and we have to identify the latter position. While we can use the print debugging, we adopt a similar but a different approach, inserting a macro "assert". The macro "assert" is a macro asserting that the condition expressed by the argument expression should be satisfied. If the condition is not satisfied, then the macro "assert" shows some messages and terminates the program.

```
#include <assert.h>
double func(int index){
  assert(index < N); // <- The program terminates if this
      condition is not satisfied.
  return data[index];
}
```

By inserting the above macro "assert" before the position where the value of index changes.[3] If the conditions written in the arguments of the macro "assert" is not satisfied, then the program terminates with the following message.

```
Assertion failed: (i<10), function func, file test.
    cpp, line 7.
```

We can find the point at which the invalid value is set to the variable.

Suppose the position of the problem is the following code.

```
#include <stdlib.h>
double myrand_double (void){
return (double)(rand())/(double) (RAND_MAX);
}
int myrand_int (const int N){
return (int)(myrand_double()*N);
}
```

The function rand returns a pseudo-random number in the range of 0 to RAND_MAX. Therefore, casting the returned value of rand to a floating point number (double) and dividing by RAND_MAX, we would have a uniform random number in the range of 0 to 1. Then we would have a random number in the range of 0 to $N-1$ by multiplying N and casting the value to an integer. This is the aim of the above program. Can you immediately identify why this program causes a bug? Actually, the program almost works as intended, and therefore, the author overlooked the bug. Because the rand returns the value from 0 to RAND_MAX uniformly and RAND_MAX is 2147483647, the rand returns RAND_MAX with the probability of two billionth in one. The function myrand returns N if and only if such a thing happens. Because myrand was expected to return a less than N, unexpected behavior occurs. While I performed a unit test for myrand when I developed this code, I did not notice this bug because the test called myrand only ten thousand times. While it is difficult to

---

[3]You may use watchpoints of debugger for this purpose.

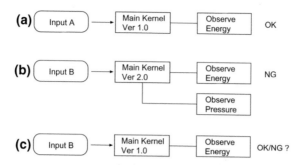

Fig. 9.3 When you find a bug, you know the following two facts: **a** The previous program does not terminate for Input A. **b** The current program terminates for Input B. Then, retrieve the previous version from the repository, and give Input B to it, and see what happens. **c** If it terminates, then the bug exists in the previous version, otherwise, the bug is inserted by the modification from the previous to the current version

identify such kinds of bugs immediately, we can surely find them with the help of printf and "assert".

9.3.2 Debugging Using Version Control Systems

In this section, we introduce several debugging techniques using version control systems. While we take Subversion as an example, you can use other software such as Git.

Suppose you add a routine calculating pressure to an MD code. In addition to this, you modify the time evolution routines. Then you find that the program terminates for some initial condition. Before debugging, you have to identify whether the bug exists before the modification or not. If your code is managed by the version control system, it is easy. Retrieve the previous version of the MD code from the repository, and give the initial condition which terminates the current program. If the previous version also terminates, then you find that the bug exists before the modification. If not, then you find the bug is inserted by the latest modification. See Fig. 9.3 for a schematic diagram of this procedure.

We have introduced the binary search to find the position of a bug in the above. We can adopt the binary search also on the time axis if your source codes are under the version control. See Fig. 9.4. The horizontal and vertical axes are the time of development and debugging, respectively. Suppose you find a bug at Rev. 5. But you should not start debugging at this moment. You first have to identify whether the bug is reproducible. Next, you should find out when the bug is inserted by means of the binary search along the revision number, i.e., the development history. Then you find out that Rev. 2 does not contain the bug while Rev. 3 dose. Now, you are ready to start debugging. Use svn diff to see the difference between two revisions. The bug is there.

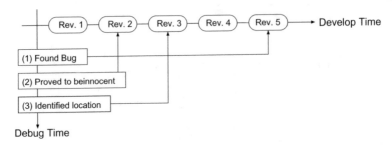

Fig. 9.4 Development and debugging procedures. **a** You find a bug at Rev. 5. **b** You identify the point at which the bug was inserted by the binary search and you confirm that Rev. 2 is the latest version which does not contain the bug. **c** Now you find that the bug was inserted at Rev. 3

The above procedures are summarized as follows:

1. Find a bug.
2. Check whether the bug is reproducible.
3. Find the latest revision without the bug by the binary search along the development history.
4. See the difference between the version with and without the bug.

Note that the above procedure can be performed in an automatic manner without using your brain.

9.3.3 Summary

It would be wonderful if you develop a code without any bugs, but in reality, bugs are inevitable. Sometimes one rushes to debug when a bug appears, but this is a bad move. Usually, you are tired when you find a bug. Debugging is unproductive, but it is highly intellectual activity at the same time. It is better not to get started on such a difficult task when you are tired. If you find a bug, take a rest after checking the reproducibility, and fix it tomorrow if possible. On debugging, do not use your brain as possible. While it is helpful to use a debugger and/or IDE, the most important thing of debugging is to obtain valuable information mindlessly. Version control systems, such as Subversion and Git, will be helpful for debugging. If you have not used the version control systems, it is a good opportunity to start using them.

9.4 How to Use Profiler

In the numerical simulation, the most time-consuming process is debugging. You may not try to optimize your code before debugging your code. If you are sure that your code is bug-free, or you can fix bugs immediately, then you are ready to optimize

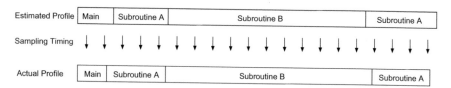

Fig. 9.5 Sampling of execution time by means of a profiler. The profiler probes the program every constant time. The arrows denote the timings of probes. Since the probe intervals are finite, estimation error is unavoidable. The bottom is an actual profile and the top is estimated one. The estimation error becomes more significant for larger probe intervals

your code. In optimization, what you have to do first is observation, not optimization. While a program usually consists of many routines, only very few routines are time-consuming. You should determine which part of the codes is the bottleneck. The most time-consuming part is called the hotspot. Although you can make a guess the position of the hotspot, you should analyze your code using some tools. There may be a hidden hotspot that you have not noticed. The tool which identifies the hotspot of the program is called a profiler. In this section, we introduce the use of it.

9.4.1 What is a Profiler?

A profiler is a tool which analyses the profile of the target program. While there are many kinds of tools to analyze a code, we introduce two of them, a statistical profiler and an event-based profiler.

9.4.2 Statistical Profiler

The statistical profiler analyzes a program by means of sampling of execution time. The profiler probes the current status of the executing program at regular intervals, and estimates the execution time of each routine. One of the popular samplers is gprof. If you want to use gprof, you have to recompile your code with an appropriate option. For example, the option -pg is required for GCC. With such option, the compiler inserts timing function before and after each routine of the code. When the code runs, the code also probes the currently executing routine at regular intervals. Suppose the time interval is 0.01 s. When some routine is observed at a timing of a probe, then the execution time of the routine can be from 0 to 0.02 s. But the profiler assumes that the routine executes for 0.01 s (see Fig. 9.5). The profile estimation always includes such estimation errors. Usually, such an estimation error is negligible. However, it becomes significant when the program calls small routine many times. In such a case, you should be aware of the time intervals of profiling.

List 9.4 Sample code for profiling

```cpp
#include <iostream>
const int N = 10000;
double x[N] = {};
double y[N] = {};
double a[N][N] = {};

void
matvec(void) {
  for (int i = 0; i < N; i++) {
    for (int j = 0; j < N; j++) {
      y[i] += a[i][j] * x[j];
    }
  }
}

double
vecvec(void) {
  double sum = 0.0;
  for (int i = 0; i < N; i++) {
    sum += x[i] * y[i];
  }
}

int
main(void) {
  for (int i = 0; i < 1000; i++) {
    vecvec();
  }
  matvec();
}
```

List 9.4 is a sample code which calculates vector–vector and vector–matrix products. Suppose N is a dimension of vectors and matrices. The complexities of computations of vector–vector and vector–matrix products are $O(N)$ and $O(N^2)$, respectively. Therefore, the vector-matrix product is expected to be a hotspot for larger N. First, compile the code with an appropriate option. We use GCC, and the option is –pg.

```
$ g++ -pg prof.cpp
$ time ./a.out
./a.out 0.32s user 0.18s system 99% cpu 0.512 total

$ ls
a.out* gmon.out prof.cpp
```

After execution, a file gmon.out is generated. Giving the execution binary and the gmon.out to gprof, gprof gives you the profile of the program. The results of gprof

```
$ gprof -bp

Flat profile:

Each sample counts as 0.01 seconds.
  %   cumulative   self              self     total
 time   seconds   seconds    calls  ms/call  ms/call  name
93.88     0.46      0.46        1   460.00   460.00  matvec()
 6.12     0.49      0.03     1000     0.03     0.03  vecvec()
 0.00     0.49      0.00        1     0.00     0.00  _GLOBAL__sub_I_x
```

Fig. 9.6 The result of gprof. This is a flat profile. The names of routines are listed in descending order, i.e., the most time-consuming routine is placed at the top

is shown in Fig. 9.6.[4] Here the options -b and -p are the options which suppress the verbose blurbs and prints only a flat profile. The flat profile shows the total amount of time of each routine in the target program. Refer the details by executing man gprof.

First, you can see the sentence "Each sample counts as 0.01 s." From this, you see that the sampling was done every 0.01 s. Next, "% time" denotes the percentage of the total execution time the program spent in this function. You can find that over 90% of execution time is spent by matvec, and therefore, matvec is the hotspot. The estimated computation time of vecvec is 0.03 s, but the value is not reliable because it is not sufficiently longer than the sampling interval 0.01 s. As a rough guideline, if the sampling is 100 or more, in this case, the execution time is more than 1 s, the execution time is generally reliable.

While gprof is useful and popular, more functional tools appear recently. Here, we introduce perf which is helpful to analyze the codes. The profiler perf is available on Linux based systems. The usage of perf is simple. You do not have to recompile your code. Compile List 9.4 without options and execute perf record a.out. After execution, the command perf report gives the profiles of your codes. See List 9.5 for usage of perf.

You can find that matvec spent about 60% and vecvec spent about 6% in the total execution time, respectively. Since perf cooperates with the Linux kernel, we can see the kernel information such as memory allocation.

List 9.6 shows the profile of the MD codes of the author. You can see that the force calculation spent 46% of the total time. The pair list construction routine follows. Note that the system size of the run is small for the test. For a product run, the force calculation becomes more significant and spends over 80% of the execution time.

Usually, a few routines spend most of the execution time of the program. Then you can improve the performance by optimizing the routines. However, a program sometimes consists of many routines which equally spend time. In such cases, the optimization is ineffective. For example, suppose there exists a routine which spent 95% of the execution time. If you halve the execution time of the routine, then you will obtain 1.9 times better performance in total. But the performance gain will be

---

[4]The actual result contains some routines such as __static_initialization_and_ destruction_0(int, int) which is automatically added routine by g++, but they are omitted.

List 9.5 How to use perf

```
$ g++ prof.cpp

$ perf record ./a.out
[ perf record: Woken up 1 times to write data ]
[ perf record: Captured and wrote 0.266 MB perf.data
   (~11642 samples) ]

$ perf report | head
# Events: 596 cycles
#
# Overhead Command Shared Object Symbol
# ........ ....... ..................
       ..............................
#
  60.13% a.out a.out [.] matvec()
   8.93% a.out [kernel.kallsyms] [k] page_fault
   6.08% a.out a.out [.] vecvec()
   2.63% a.out [kernel.kallsyms] [k]
       isolate_freepages_block
   1.47% a.out [kernel.kallsyms] [k] isolate_migratepages
```

List 9.6 Profile of an MD code

```
# Events: 30K cycles
#
# Overhead Command Shared Object Symbol
# ........ ....... ..................
#
  46.75% a.out a.out [.] MD::calculate_force_list()
  38.99% a.out a.out [.] MeshList::search_other()
   4.39% a.out a.out [.] Observer::potential_energy()
   3.36% a.out a.out [.] Variables::make_neighbor_list()
   2.00% a.out a.out [.] MD::run()
```

only 6.9 times even if you optimize the routine tenfold. The gain from optimization becomes smaller and smaller as it takes time and effort to optimize. It is important to determine a rough target performance before optimizing the program.

9.4.3 Event-Based Profilers

Recent CPUs have hardware counters which allow us to count the various events, such as instruction per cycles (IPC), a number of branch instructions, a number of branch-misses, and so forth. Using a profiler which can access such hardware counters, we can analyze the performance of the code in detail.

On Linux based systems, we can obtain the information of events with perf. Just run `perf stat ./a.out`. Then you will obtain the following results.

```
$ perf stat ./a.out

       5587.039655 task-clock # 0.997 CPUs utilized
               550 context-switches # 0.000 M/sec
                 2 CPU-migrations # 0.000 M/sec
           195,647 page-faults # 0.035 M/sec
    16,170,296,845 cycles # 2.894 GHz
       <not counted> stalled-cycles-frontend
       <not counted> stalled-cycles-backend
    19,770,941,849 instructions # 1.22 insns per cycle
     3,103,214,731 branches # 555.431 M/sec
        15,888,704 branch-misses # 0.51% of all
                   branches

       5.603996583 seconds time elapsed
```

The sentence "0.997 CPUs utilized" means that about one CPU-core is used for the execution. If you run the multithread program, then the value becomes larger. The sentence "1.22 insns per cycle" denotes the average number of IPC. The small IPC suggests that CPU is almost idle. "branches" and "branch-misses" denote the number of branches and the number of misses. The branch usually appears as a conditional jump such as "jump to some address when this variable is true, otherwise jump to another address". The recent CPUs usually have instruction pipelines and the depth becomes deeper, and the branch causes a serious pipeline hazard which results in a significant performance loss.[5] Therefore, recent architecture usually has a branch predictor. The branch predictor first predicts the results of the branch, then executes the following instructions on the basis of the prediction. If the prediction is incorrect, then CPU disposes of executed results. In the present case, most of the branches are correctly predicted, and therefore, the branch-misses does not contribute to the performance loss. Perf can obtain the rate of a cache-miss. You can find the list of observable events by `perf list`.

From the obtained events, we have to identify the bottleneck of the program. While there is no silver bullet, here are the summary of the typical bottlenecks and how to deal with them.

Cache-miss The high rate of cache-miss suggests that memory usage is inefficient. If the data required by CPU are not on the cache, CPU should wait a long time for fetching data from memory. This can be resolved by rewriting a program to be cache-aware. But it is difficult to deal with it when the computational task is light compared to the amount of the data.

---

[5]Refer to the previous chapter for instruction pipelines and pipeline hazards.

Barrier synchronization A multithreaded program includes many threads. There is a point in the program where all threads should stop until all threads have arrived at that point. Therefore, the programmer calls a barrier in order to synchronize all threads. If the tasks assigned to the threads are imbalanced, then some thread should wait for other threads at the barrier. This is the waiting time due to the barrier synchronization. If the waiting time for the barrier synchronization is time-consuming, then the load imbalance can be the principal cause.

Data dependency If an instruction uses data which is the results of the previous instruction, the current instruction cannot be executed until the previous instruction completes. This is called data dependency. Suppose we want to calculate "A = B + C; C = A * E;". We cannot calculate the latter until the former is completed. Even if the computational cost increases slightly, we can reduce the total performance loss by adopting algorithms with less data dependency.

Vectorization rate The performance improvement of the recent architecture depends mainly on the increase of CPU-core number and SIMD width. Because it is usually difficult to increase the vectorization ratio by hands, it would be better for you to leave it even if you find that your codes have poor vectorization ratio. If you really want to increase the vectorization ratio, then you have to write your hotspot by using intrinsic functions.

9.5 Specific Examples of Optimization

The most important thing in optimization is the complexity of the algorithm. It is meaningless to optimize an algorithm of $O(N^2)$ if there is an algorithm of $O(N)$. You can start optimizing your code when you are sure that the algorithm of your code has fixed. There are no general recipes for optimization. In the present section, we introduce several examples of optimization.

9.5.1 Cache-Aware Optimization

Recently, memory access is too slow compared with computational power. In order to address this problem, there are several caches between CPU and memory. A cache is a temporal storage of data. It can serve data to CPU faster than memory, while the capacity is much smaller than memory. If data required by CPU are not on a cache, then CPU should wait typically several hundred cycles. Therefore, cache-aware optimization is highly important to the recent architecture. While some of the cache-aware optimization techniques are described in the previous chapter, we introduce one example for MD simulations.

As time evolution, the data of the interacting atoms which are spatially close may be separated in the memory. This severely decreases the computational efficiency

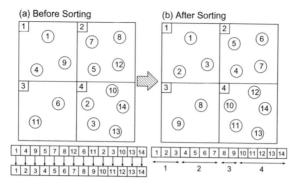

Fig. 9.7 Spatial sorting. **a** Before sorting. The atoms 1 and 9 are spatially close to each other, but they are separated on the memory. **b** After sorting. The system is divided into small cells and the atoms are sorted by the labels of cells. Then the data of the atoms which are spatially close are also sorted and placed close to each other in the memory

due to the increase of cache-miss. In order to improve the cache efficiency, we sort indices of the atoms so that the data of the atoms are also sorted in the memory and placed close to each other [5]. See Fig. 9.7 for implementation.

The computational speeds with and without spatial sorting are shown in Fig. 9.8. In order to see the efficiency of the sorting, the indices of the atoms are shuffled randomly at the beginning of the simulations. The computational speed is measured in the unit of MUPS (millions update per second), which is unity when one million atoms are updated in 1 s. The simulations were performed on a Xeon 2.93 GHz machine with 256 KB for an L2 cache and 8MB for an L3 cache. The system is three dimensional, and therefore, the information of the atoms consists of six components, three for coordinates and three for momenta. Because the double precision number is 8 bytes, one atom is expressed with 48 bytes. Therefore, the L2 cache can contain the information of 5.3×10^3 atoms and the L3 can contain that of 1.7×10^5 atoms. The open circles denote the results without sorting. As the number of atoms spills out of the capacity of each cache, the efficiency is degraded at the number of atoms mentioned above. The solid circles denote the results with spatial sorting. While the total number of atoms is larger than the capacity of caches, the computational speed is almost independent of the number of atoms. This behavior implies that the program can utilize cache efficiently.

While the optimization techniques strongly depend on the architecture on which the program runs, the cache-aware optimization is usually effective on various kinds of machines. When the performance of your code decreases significantly for larger system size, it is better to consider applying the cache-aware optimization.

9.5.2   Register-Aware Optimization

As described in the previous chapter, the storage of computers form a hierarchy. The faster storage has a smaller capacity. A processor register, which is the fastest and

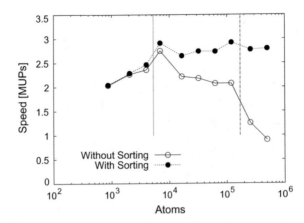

Fig. 9.8 Efficiency of spatial sorting. The open and solid circles denote results with and without sorting, respectively. The positions of dashed lines reflect the capacity of L2 and L3 caches

smallest storage, is placed at the end of the hierarchy. All data should be loaded on the register in order to perform arithmetic operations. Some program can exhibit better performance by adopting register-aware optimization. Let us see a simple example.

List 9.7 is the code which calculates the forces between all atoms. Note that, the complexity of this code is $O(N^2)$. This program has double loops. The loop variables are i and j. In order to calculate the force between i- and j-atoms, the coordinates of them are required. Now, you can see that the coordinates of i-atoms are fixed in the inner loop, and therefore, it is inefficient to fetch the data of the coordinates of the i-atoms every time from the memory. In order to reduce memory access, we introduce temporary variables. We store the data of the coordinates of the i-atoms to the local variables and reuse them in the inner-loop. We also use the temporary variables to accumulate the impulse of the i-atoms and write back them to the momenta when the inner-loop completes.

List 9.8 is the program modified in the above policy. We expect that the data of both coordinates and momenta of i-atoms are placed on the register.

The two above programs are compiled as follows:

```
$ g++ -O3 -mavx force1.cpp -o a.out
$ g++ -O3 -mavx force2.cpp -o b.out
```

The computational time is measured on the machine of Xeon E5-2680 v3 (2.50GHz). The modified program takes 1.60 s while the original does 1.86 s. Optimization usually spoils the readability of the program. Therefore, the optimization is a trade-off between readability and performance. In the present case, about 15% better performance is obtained by the simple modification, and therefore, it is worth to apply.

List 9.7 Simple force calculation (force1.cpp)

```cpp
#include <stdio.h>
enum {X, Y, Z};
const int N = 20000;
const double dt = 0.01;
double q[N][3] = {};
double p[N][3] = {};

void
init(void) {
  for (int i = 0; i < N; i++) {
    q[i][X] = 1.0 + 0.4 * i;
    q[i][Y] = 2.0 + 0.5 * i;
    q[i][Z] = 3.0 + 0.6 * i;
    p[i][X] = 0.0;
    p[i][Y] = 0.0;
    p[i][Z] = 0.0;
  }
}

void
calc(void) {
  for (int i = 0; i < N; i++) {
    for (int j = i + 1; j < N; j++) {
      const double dx = q[j][X] - q[i][X];
      const double dy = q[j][Y] - q[i][Y];
      const double dz = q[j][Z] - q[i][Z];
      const double r2 = (dx * dx + dy * dy + dz * dz);
      double r6 = r2 * r2 * r2;
      double df = (24.0 * r6 - 48.0) / (r6 * r6 * r2) * dt;
      p[i][X] += df * dx;
      p[i][Y] += df * dy;
      p[i][Z] += df * dz;
      p[j][X] -= df * dx;
      p[j][Y] -= df * dy;
      p[j][Z] -= df * dz;
    }
  }
}

int
main(void) {
  init();
  calc();
  for (int i = 0; i < 10; i++) {
    printf("%f %f %f\n", p[i][X], p[i][Y], p[i][Z]);
  }
}
```

List 9.8 Register aware version (force2.cpp)

```
void
calc(void) {
  for (int i = 0; i < N; i++) {
    const double qix = q[i][X];
    const double qiy = q[i][Y];
    const double qiz = q[i][Z];
    double pix = p[i][X];
    double piy = p[i][Y];
    double piz = p[i][Z];
    for (int j = i + 1; j < N; j++) {
      const double dx = q[j][X] - qix;
      const double dy = q[j][Y] - qiy;
      const double dz = q[j][Z] - qiz;
      const double r2 = (dx * dx + dy * dy + dz * dz);
      double r6 = r2 * r2 * r2;
      double df = (24.0 * r6 - 48.0) / (r6 * r6 * r2) * dt;
      pix += df * dx;
      piy += df * dy;
      piz += df * dz;
      p[j][X] -= df * dx;
      p[j][Y] -= df * dy;
      p[j][Z] -= df * dz;
    }
    p[i][X] = pix;
    p[i][Y] = piy;
    p[i][Z] = piz;
  }
}
```

Note that, some compiler can perform such optimization automatically. If we use Intel Compiler, then the original and modified code take 0.875 and 0.881 s, respectively. The efficiency of optimization strongly depends on not only the architecture of the machine, but also on the optimization ability of compiler.

9.6 Summary

In the present chapter, we have introduced several debugging and optimizations techniques. While people tend to focus on the execution time of a program, the time of development is usually longer, and therefore, is more important. The appropriate debugging techniques reduce the time of development drastically. The version control systems are also helpful for this purpose. You may consider optimization after we are sure that the debugging process has completed and the algorithm of the program is definitely fixed. We have to decide how seriously we optimize our code on a case-by-case basis. While the cache-aware tuning is usually effective on the wide range of computers, the vectorization using the SIMD instruction strongly depends on the architecture and lacks portability. Additionally, the performance improvement by SIMD vectorization is usually less effective than that by the cache-aware tuning and it often does not compensate the time spent for it. After cache-aware tuning, we would better to focus on producing results, rather than insisting on optimization. We would be glad if this chapter is helpful to reduce the development and execution time of your codes.

Exercises

1. Compile List 9.7 and List 9.8, and analyze them with perf. For example, find the success rate of branch predictions by running `perf stat ./a.out`. Try identical runs several times to see the fluctuation of results. Compare the results with a compile options `g++ -O3` and `g++ -O3 -mavx`.
2. Save List 9.7 as `force1.cpp` and run `g++ -O3 -S force1.cpp`. Then you will find an assembler code `force1.s`. Check whether the assembler code contains an instruction `vmulsd` or not. Do the same things by adding `-mavx` option. If Intel Compiler is available, run `icpc -O3 -xHOST -S force1.cpp` and see whether the generated assembler code `force1.s` contains the instructions such as `vmulpd` and/or `vfmadd213pd`. Investigate the roles of these instructions, too.

References

1. M.P. Allen, D.J. Tildesley, *Computer Simulation of Liquids* (Clarendon Press, Oxford, 1987)
2. K. Beck, *Test Driven Development: By Example, Boston* (Addison-Wesley, MA, 2003)
3. K. Beck, *Extreme Programming Explained: Embrace Change* (Addison-Wesley, Boston, MA, 2000)
4. J. Rasmusson, *The Agile Samurai: How Agile Masters Deliver Great Software* (Pragmatic Programmers, 2010)
5. H. Watanabe, M. Suzuki, N. Ito, Prog. Theor. Phys. **126**, 203–235 (2011)

Chapter 10
Techniques Concerning Computation Accuracy

Shin'chi Oishi, Yusuke Morikura, Kouta Sekine, Hisayasu Kuroda and Maho Nakata

Abstract This chapter deals with fundamental theories on the accuracy of numerical calculation and some cases that seems to be important, somewhat different from previous chapters. We must remember that numerical errors are included in the output data of the computer. In particular, do not overlook the important points you need to know when parallelizing codes. Pursuit of calculation speed is, of course, the central theme of this book, however, it is premised that it produces correct results. This chapter introduces a numerical computation method with guaranteed accuracy in large-scale numerical computations, convergence accuracy problems in parallel computing, and high-precision calculation in HPC.

S. Oishi (✉)
Faculty of Science and Engineering, Waseda University, Shinjuku, Tokyo, Japan
e-mail: oishi@waseda.jp

Y. Morikura
Faculty of Modern Life, Teikyo Heisei University, Nakano, Tokyo, Japan
e-mail: y.morikura@thu.ac.jp

K. Sekine
Department of Information Networking for Innovation and Design, Toyo University,
Kita, Tokyo, Japan
e-mail: k.sekine@computation.jp

H. Kuroda
Graduate School of Science and Engineering, Ehime University, Matsuyama, Ehime, Japan
e-mail: kuroda@cs.ehime-u.ac.jp

M. Nakata
Head Office for Information Systems and Cybersecurity, RIKEN, Wako, Saitama, Japan
e-mail: maho@riken.jp

M. Geshi (ed.), *The Art of High Performance Computing for Computational Science, Vol. 1*, https://doi.org/10.1007/978-981-13-6194-4_10

10.1 A Numerical Computation Method with Guaranteed Accuracy in Large-Scale Numerical Computations

The current computer performs mathematical arithmetic that approximates floating-point numbers stipulated in the IEEE 754 Standard [1]. When the result of an arithmetic using only floating-point numbers cannot be expressed as a floating-point number, the result is converted into a floating point number through approximation by rounding up/down. This is called a rounding error. Although this allows floating-point number arithmetic to be done fast, the results obtained through such arithmetic are approximate values that have rounding errors. In addition to rounding errors, discretization error occurs in numerical integration or differential equations, and truncation errors occur in Taylor expansion or Newton's method. In this sense, results obtained from a computer are not necessarily correct. Therefore, there is a need for a method to check the accuracy of the results obtained with approximate values; furthermore, once the accuracy has been checked, there is also sometimes a need to improve the approximate values. This section introduces methods that guarantee computation accuracy in large-scale numerical computations.

There is a method for guaranteeing the accuracy of numerical computations that serves as a means to check the accuracy of the results when obtained with approximate values that include errors. Numerical computation with guaranteed accuracy is a method that obtains a mathematically strict bound for errors in the exact solution x^* and approximate solution $\hat{x}$ for a problem to be solved, such as eigenvalue problems, integral equations, or differential equations, including simultaneous linear equations. Therefore, it can be used not only to estimate errors in an approximate solution, but also as a tool for mathematical verification.

This section will first discuss interval arithmetic, the underlying approach of numerical computation with guaranteed accuracy. As an example, a method of numerical computation with guaranteed accuracy with regard to simultaneous linear equations will be discussed. For details on numerical computation with guaranteed accuracy in other problems, please refer to references [2, 3].

10.1.1 Rounding and Interval Arithmetic in Floating-Point Number Arithmetic

Let $\mathbb{F}$ be a set of floating-point numbers stipulated in the IEEE 754 Standard. Here, it is assumed that no overflow will occur. The four basic arithmetic operations are $\circ \in \{+, -, \times, /\}$. In the IEEE 754 Standard, results are approximated to five arithmetic—the four basic arithmetic operations used with floating-point numbers and the square root operation—and the standard stipulates the mode for rounding, such as rounding up, rounding down, and rounding to nearest. For example, for $a, b \in \mathbb{F}$, the four basic arithmetic operations are

- Round up: $\overline{\circ} : \mathbb{F} \times \mathbb{F} \to \mathbb{F}$ as inf $\{x \in \mathbb{F} \mid x \geq a \circ b\}$
- Round down: $\underline{\circ} : \mathbb{F} \times \mathbb{F} \to \mathbb{F}$ as sup $\{x \in \mathbb{F} \mid x \leq a \circ b\}$
- Round-to-nearest: $\tilde{\circ} : \mathbb{F} \times \mathbb{F} \to \mathbb{F}$ as rounded to the closest floating-point number.[1]

For example, using **fenv.h** in a C99-conformant C language compiler allows you to use the **fesetround** function, which changes the rounding mode.[2,3]

Next, we will explain interval arithmetic, a basic tool for estimating rounding errors. Let $\mathbb{R}$ be a set of real numbers. An interval with a floating-point number at its endpoint is defined as $[a, b] := \{x \in \mathbb{R} \mid a \leq x \leq b, \ a, b \in \mathbb{F}\}$. First, the four basic interval arithmetic are applied to the two intervals $[a_l, a_u], [b_l, b_u]$. Basic interval arithmetic estimates the interval $[c_l, c_u]$ that includes the range $f([a_l, a_u], [b_l, b_u]) := \{z \in \mathbb{R} \mid z = f(x, y), \ x \in [a_l, a_u], \ y \in [b_l, b_u]\}$ for the four basic arithmetic operations $f : \mathbb{R} \times \mathbb{R} \to \mathbb{R}$. The interval arithmetic in the interval $[a_l, a_u]$ and $[b_l, b_u]$ can be calculated, for example, as follows:

- Addition: $[a_l, a_u] + [b_l, b_u] = [a_l \underline{+} b_l, a_u \overline{+} b_u]$
- Subtraction: $[a_l, a_u] - [b_l, b_u] = [a_l \underline{-} b_u, a_u \overline{-} b_l]$
- Multiplication: $[a_l, a_u] \times [b_l, b_u] = [\min\{a_l \underline{\times} b_l, a_u \underline{\times} b_l, a_l \underline{\times} b_u, a_u \underline{\times} b_u\},$
  $\max\{a_l \overline{\times} b_l, a_u \overline{\times} b_l, a_l \overline{\times} b_u, a_u \overline{\times} b_u\}]$
- Division: $[a_l, a_u]/[b_l, b_u] = [\min\{a_l \underline{/} b_l, a_u \underline{/} b_l, a_l \underline{/} b_u, a_u \underline{/} b_u\},$
  $\max\{a_l \overline{/} b_l, a_u \overline{/} b_l, a_l \overline{/} b_u, a_u \overline{/} b_u\}]$, where $0 \notin [b_l, b_u]$.

Multiplication and division can be handled separately based on the plus or minus sign at the endpoint.

10.1.2 A Numerical Computation Method with Guaranteed Accuracy for Large-Scale Simultaneous Linear Equations

This section will discuss a method of numerical computation with guaranteed accuracy as applied to the approximate solution $\hat{x}$ of the simultaneous linear equation

---

[1] For details on specifications in which there are two closest floating-point numbers or an arithmetic is approaching overflow, please consult the standard [1].

[2] In Fortran language, when in conformity with the Fortran2003 standard, modules **IEEE_ARITHMETIC** or **IEEE_FEATURES** make it possible to change the rounding mode. For instance, by inputting the text **CALL IEEE_SET_ROUNDING_MODE(IEEE_NEAREST)**, round-to-nearest will be the selected mode. The mode will be changed to rounding up if **IEEE_NEAREST** is replaced with **IEEE_UP**, and changed to rounding down if replaced with **IEEE_DOWN**.

[3] If the computation order is changed owing to compiler optimization, an operation not conforming to the IEEE 754 Standard may be performed, unintentionally resulting in an operation that does not constitute a numerical computation with guaranteed accuracy. Therefore, in order to inhibit optimization, it is necessary to add a **volatile** attribute stipulated in C language and Fortran2003 standards as a variable, or to set up arithmetic so that they more strictly conform if optimization options (**-fp-model** etc.) for floating-point numbers are included as compiler options.

$$Ax = b, \ A \in \mathbb{F}^{n \times n}, \ b \in \mathbb{F}^n. \tag{10.1}$$

The method of numerical computation with guaranteed accuracy precisely estimates the error bound for the approximate solution $\hat{x}$ and the exact solution x^*. For example, the theorem below is often used in numerical computations with guaranteed accuracy for simultaneous linear equations.

Theorem 10.1 *Let* A, $R \in \mathbb{R}^{n \times n}$, *and* $\hat{x}$, $b \in \mathbb{R}^n$. I *is an identity matrix. If*

$$\|RA - I\| < 1 \tag{10.2}$$

is satisfied, A *is invertible and*

$$\|\hat{x} - A^{-1}b\| \le \frac{\|R(A\hat{x} - b)\|}{1 - \|RA - I\|} \tag{10.3}$$

is satisfied.

In a computation, $\hat{x} \in \mathbb{F}^n$ is the approximate solution to Eq. (10.1), $R \in \mathbb{F}^{n \times n}$ is the approximate inverse matrix of A, and the norm is evaluated using the maximum-value norm. The maximum-value norm in $x \in \mathbb{R}^n$, $A \in \mathbb{R}^{m \times n}$ is defined as

$$\|x\|_\infty := \max_{1 \le i \le n} |x_i|, \quad \|A\|_\infty := \max_{1 \le i \le m} \sum_{j=1}^{n} |a_{ij}|.$$

If (10.2) is satisfied, the error bound between the approximate solution and the exact solution will be obtained from (10.3). However, because the floating-point number arithmetic in (10.2) and (10.3) include rounding errors, these errors must be taken into consideration when calculating the upper bound. We must use the interval arithmetic from Sect. 10.1.1. Because optimization is not performed when the matrix product of RA and the like is calculated using the interval arithmetic from Sect. 10.1.1, the execution speed becomes very slow. In order to reduce the number of interval arithmetic, the right numerator of (10.3) is as

$$\|R(A\hat{x} - b)\| \le \|R\| \|A\hat{x} - b\|. \tag{10.4}$$

Although this makes the evaluation more complicated, it allows the evaluation to be completed without introducing a point matrix and the product of an interval vector.

Furthermore, using the optimized **BLAS**, it is possible to include the exact value in the matrix product RA and the matrix–vector product $A\hat{x}$. This method of inclusion will be introduced next.

10.1.3 Fast Inclusion Method for a Matrix Product Using the Switches of Rounding Mode

This subsection describes a fast inclusion method for the matrix product RA for $R, A \in \mathbb{F}^{n \times n}$ [4]. It can be calculated as follows:

```
#include <fenv.h>
fesetround(FE_UPWARD);        //Change to rounding up
C_u=R*A;                      //Matrix product with rounding up
fesetround(FE_DOWNWARD);      //Change to rounding down
C_d=R*A; //Matrix product with rounding down
```

By using **BLAS** optimized for the matrix product $R * A$, we can fast obtain inclusion for the matrix product $R * A$. However, under general matrix products, such as with the product of the three matrices $A * B * C$, we must be aware that the inclusion of results is not possible because of an interval matrix arithmetic. Furthermore, although the rounding mode is changed before performing the matrix product $R * A$, the rounding mode must be changed for all nodes and threads used in matrix arithmetic. Therefore, a guarantee that the rounding mode of all nodes and threads used in **BLAS** have been changed becomes necessary. Furthermore, because subtraction is included in algorithms that reduce computational complexity, such as the Strassen algorithm, inclusion of the matrix product is not possible even if the rounding mode for all matrix product is rounding up (or rounding down). Therefore, a guarantee that algorithms such as the Strassen algorithm have not been used for **BLAS** is required. Based on this point, the next section will introduce a method that uses only round-to-nearest. When algorithms such as the Strassen algorithm are not used in **BLAS**, the method of using only round-to-nearest is an effective technique that can be used in situations where rounding changes are not guaranteed.

10.1.4 Numerical Computation Method with Guaranteed Accuracy of Solutions to Simultaneous Linear Equations Using Only Round-to-Nearest

We must pay attention to whether the rounding mode has been properly changed (in all nodes and all threads) in parallel environments. Therefore, a numerical computation method [5] has been proposed that uses **BLAS** in the solutions of simultaneous linear equations by obtaining rounding errors with round-to-nearest and a priori estimate [6]:

Theorem 10.2 (Ogita et al. [5]) *Let $A \in \mathbb{F}^{n \times n}$, and $b \in \mathbb{F}^n$ be given. Let $\hat{x} \in \mathbb{F}^n$ be an approximate solution of the simultaneous linear equation $Ax = b$, and $R \in \mathbb{F}^{n \times n}$ be an approximate inverse matrix of A. Let $e = (1, 1, \ldots, 1)^T \in \mathbb{F}^n$. $\mathbf{u}$ is a relative rounding error unit and $\mathbf{u}_N$ is the smallest positive normalized number in a given* *working precision (e.g.* $\mathbf{u} = 2^{-53}$ *and* $\mathbf{u}_N = 2^{-1022}$ *for double-precision floating-point numbers in the IEEE 754 Standard).* $\mathrm{fl}(\cdots)$ *denotes that each operation in the parenthesis is evaluated by floating-point number arithmetic with rounding to nearest.* $\tilde{\gamma}_n = \mathrm{fl}((n\mathbf{u})/(1 - n\mathbf{u}))$. α_1, α_2 *is*

$$\alpha_1 := \mathrm{fl}\left(\|RA - I\|_\infty\right), \quad \alpha_2 := \mathrm{fl}\left(\||R|(|A|e)\|_\infty\right),$$

where $|\cdot|$ *expresses the absolute value of each element. If both* $(3n + 2)\mathbf{u} < 1$ *and* $\alpha_1 < 1$ *are satisfied, the upper bound of* $\|RA - I\|_\infty$ *can be evaluated as*

$$\|RA - I\|_\infty \le \mathrm{fl}\left(\frac{\alpha_1 + \tilde{\gamma}_{3n+2}(\alpha_2 + 2)}{1 - 2\mathbf{u}}\right) =: \alpha.$$

Here, if $\alpha < 1$, *then*

$$r_{\mathrm{mid}} := \mathrm{fl}\left(A\hat{x} - b\right), \quad r_{\mathrm{rad}} := \mathrm{fl}\left(\tilde{\gamma}_{2n+4}\{(|A||\hat{x}| + |b|) + \mathbf{u}^{-1}\mathbf{u}_N \cdot e\}\right)$$

is satisfied. t and q are

$$t := \mathrm{fl}\left(\tilde{\gamma}_{n+1}\max(|r_{\mathrm{mid}}|, \mathbf{u}_N \cdot e)\right), \quad q := \mathrm{fl}\left(\frac{|R|(t + r_{\mathrm{rad}}) + 2\mathbf{u}_N \cdot e}{1 - (n + 3)\mathbf{u}}\right),$$

where $\max(\cdot, \cdot)$ *is an expression of a large element obtained by comparing each element. Here, this forms the following inequality:*

$$\|R(A\hat{x} - b)\|_\infty \le \mathrm{fl}\left(\frac{\||R \cdot r_{\mathrm{mid}}| + q\|_\infty}{1 - 2\mathbf{u}}\right) =: \beta.$$

By using the obtained α *and* β *to calculate*

$$\|\hat{x} - A^{-1}b\|_\infty \le \mathrm{fl}\left(\left(\frac{\max(\beta, \mathbf{u}_N)}{1 - \alpha}\right)/(1 - 3\mathbf{u})\right),$$

the upper bound for $\|\hat{x} - A^{-1}b\|_\infty$ *is obtained.*

Because this theorem uses round-to-nearest only, there is no longer a need to guarantee that the rounding mode of all nodes and threads has been changed, which is required in **BLAS**. Therefore, **BLAS** can be used to calculate α_1, which has the largest computational complexity, as long as the Strassen algorithm is not used. For details on a priori error estimate, please refer to Refs. [5, 6].

10.1.5 Numerical Simulation

Last, in this section, we will show an example of a numerical simulation. The simulation was performed in the following environment:
OS: CentOS 6.5
CPU: Intel Xeon Processor E7-4830 v2 (20M Cache, 2.20 GHz), 4 CPU/40 cores
Memory: 2Tbyte (DDR 3)
Software: MATLAB2016a.

The algorithm executed was as follows:
Alg1: Using the switches of rounding mode, (10.3) in Theorem 10.1 was calculated as (10.4).
Alg2: Using only rounding-to-nearest, Theorem 10.2 was calculated.

```
function [alpha,err] = Alg1(A,R,b)
  feature('setround',0.5);
  n=length(A);
  I=eye(n);
  feature('setround',-inf);
  Gd = (abs(R*A-I));
  feature('setround',inf);
  Gu = (abs(R*A-I));
  Gu =max(Gd,Gu);
  alpha=norm(Gu,inf);
  if alpha >=1, ...
  error('verification failed.'), end
  feature('setround',0.5);
  x=R*b;
  feature('setround',-inf);
  rd = abs(A*x-b);
  alpha_=1-alpha;
  feature('setround',inf);
  ru = abs(A*x-b);
  ru = max(rd,ru);
  R_up=norm(abs(R),inf);
  ru_=norm(ru,inf);
  beta=R_up*ru_;
  err=beta/alpha_;
  feature('setround',0.5);
end
```

```
function [alpha,err] = Alg2(A,R,b)
  u=2^-53;
  u_=2^-1022;
  feature('setround',0.5);
  n=length(A);
  I=eye(n);
  e=ones(n,1);
  alpha_1 = norm(R*A-I,inf);
  alpha_2 = norm(abs(R)*(abs(A)*e),inf);
  if (3*n+2)*u >=1, ...
  error('verification failed.'), end
  if alpha_1 >=1,...
   error('verification failed.'), end
  alpha =  (alpha_1+...
  g_t(3*n+2)*(alpha_2+2))/(1-2*u);
  if alpha >=1, ...
   error('verification failed.'), end
  x=R*b;
  r_mid=A*x-b;
  r_rad=g_t(2*n+4)*(abs(A)*abs(x)...
   +abs(b)+u^-1*u_*e);
  t = g_t(n+1)*max(abs(r_mid),u_*e);
  q =  (abs(R)*(t+r_rad)+2*u_*e)/(1-(n+3)*u);
  beta=norm(abs(abs(R*r_mid)+q),inf)/(1-2*u);
  err = (max(beta,u_)/(1-alpha))/(1-3*u);
end
```

```
function res = g_t(n)
  u=2^-53;
  res=(n*u)/(1-n*u);
end
```

The simulation was performed five times for each algorithm with the matrix size n = 10,000, 20,000, and 50,000, and the averages are listed. The test matrix was created as "$A = \text{randn}(n)$, $b = A * \text{ones}(n, 1)$" in MATLAB functions. In MATLAB, the rounding mode of feature("setround", mode) was changed to rounding up for **inf**, round-to-nearest for **0.5**, and rounding down for **-inf**.

According to Table 10.1, the computational results of upper bounds of $\|RA - I\|_\infty$ is overestimated more in **Alg2**, which uses a priori error estimation, than in **Alg1**. According to the computational results of upper bounds of $\|\hat{x} - A^{-1}b\|_\infty$ shown in Table 10.2, the norm in the results of **Alg1** is overestimated as in (10.4). This demonstrates that the results of **Alg1** are more overestimated than the results of **Alg2**, which uses a priori error estimation. Additionally, even when the switches of rounding mode are used, the evaluation can be even better than that of **Alg2** if

Table 10.1 Upper bound of $\|RA - I\|_\infty$

n	Alg1	Alg2
10,000	3.1319e−07	1.7342e−05
20,000	2.1572e−07	4.2553e−04
50,000	1.7369e−06	0.0063

Table 10.2 Upper bound of $\|\hat{x} - A^{-1}b\|_\infty$

n	Alg1	Alg2
10,000	5.3998e−05	1.1549e−05
20,000	0.0013	2.8581e−04
50,000	0.0618	0.0043

Table 10.3 Computation time (s)

n	R	Alg1	Alg2
10,000	9.0790	6.6569	3.6661
20,000	50.8709	47.3164	25.2638
50,000	626.3300	710.7150	361.4290

R means the computational time of approximate inverse of A

adjustments are made to directly calculate (10.3). Lastly, because **Alg2** uses a priori error estimation to reduce the number of computations for the matrix product of (10.2), we can see from Table 10.3 that **Alg2** is faster than **Alg1**. Therefore, **Alg2** is an algorithm for supercomputers.

Because each element of the exact solution x^* is almost 1, in this test problem only, the results of Table 10.2 are considered to nearly reach the relative error bound. Because the relative error for one instance of rounding in a double-precision floating-point number arithmetic is $2^{-53}(\approx 10^{-16})$, errors in the results of Table 10.2 appear large overall. We will omit the details, but it is possible to improve approximate solutions and evaluation of error by using high-precision computations (e.g. [7]). An outline of this is shown below:

- The residual iteration $\hat{x}_{i+1} = \hat{x}_i - R(A\hat{x}_i - b)$ is calculated and the approximate solution is improved. Here, the residual $A\hat{x} - b$ is computed with high precision.
- High-precision computation with error estimation is used to compute the residual $A\hat{x} - b$ in the error bound (10.3).

Using these techniques makes it possible to reduce the final relative error bound obtained from 10^{-10} to 10^{-16}.

10.2   Convergence Accuracy Problem in Parallel Computing

In a time-lapse simulation, when an iterative solver is used within the computation and the iteration is terminated in a fixed time, the numerical error accumulates and propagates, and in many cases the result of the simulation is greatly influenced. Even if you solve it with sufficient accuracy at the iterative solver, there are some points to be aware of in parallel computing.

First, in the case of sequential processing, exactly the same simulation result can be obtained anytime. Even in the case of simulation using random numbers, exactly the same simulation result is expected by setting the same seed value (the initial value internally used by the pseudo-random number generation algorithm).

On the other hand, in parallel computing, the simulation result may be greatly different every time it runs. To make matters worse, it is not reproducible, and the same simulation result cannot be obtained again. In this section, we will discuss such cases and solutions. Although the MPS method is cited as a specific example here, it is a problem that occurs also in general parallel computing.

10.2.1   MPS Method

The Moving-Particle Semi-implicit (MPS)method is a computational method for the simulation of incompressible free surface flows developed by Koshizuka and Oka in 1995 [8]. In the particle method, the continuum to be simulated is represented by a finite number of particles, and the behavior of the continuum is calculated by the movement of the particle. In the MPS method, the governing equations of the continuum is discretized by using the intergranular interaction model corresponding to the differential operator.

Here, the particle interaction model is a computational model that if the distance between r two particles is less than or equal to the influence range r_e of the interaction, the two particles interact (Fig. 10.1).

Fig. 10.1 Particle interaction model

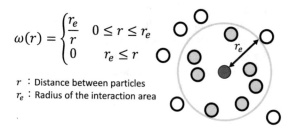

$$\omega(r) = \begin{cases} \dfrac{r_e}{r} & 0 \leq r \leq r_e \\ 0 & r_e \leq r \end{cases}$$

r : Distance between particles
r_e : Radius of the interaction area

It is necessary to search for all particles within the radius of the interaction area. (It is called Neighbor-particle searching)

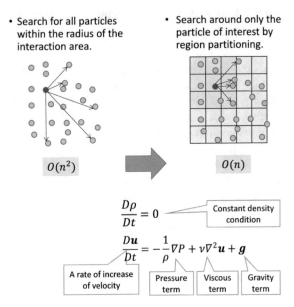

Fig. 10.2 Region partitioning in neighboring particle search

• Search for all particles within the radius of the interaction area.

• Search around only the particle of interest by region partitioning.

$O(n^2)$ $O(n)$

Fig. 10.3 Navier–Stokes equations for stationary viscous incompressible flow

$$\frac{D\rho}{Dt} = 0 \quad \text{Constant density condition}$$

$$\frac{Du}{Dt} = -\frac{1}{\rho}\nabla P + \nu\nabla^2 u + g$$

A rate of increase of velocity   Pressure term   Viscous term   Gravity term

In the particle interaction model, it is necessary to search for all possible pairs of particles within an interaction radius. Assume that the number of particles is n, the computational complexity cost is $O(n^2)$ in a simple implementation, however, using region partitioning makes it possible that the computational complexity cost is close to $O(n)$ (Fig. 10.2).

In order to simulate the phenomenon of the fluid, Navier–Stokes equations for stationary viscous incompressible flow is used (Fig. 10.3).

In the MPS method, the pressure term is implicitly computed, and the viscosity and the gravity term are explicitly computed. It is common to use the conjugate gradient method when solving the Poisson's pressure equation in the computation of the pressure term. In the MPS method, while repeating (i) an explicit solution section, (ii) neighboring particle search, (iii) matrix generation, and (iv) implicit solution section (conjugate gradient method), a time-lapse simulation is performed.

The conjugate gradient method is one of iterative methods for solving linear equations $Ax = b$, and the algorithm is shown in Fig. 10.4.

When parallel computing is actually carried out, we often see the behavior that the result differs with each execution as shown in Fig. 10.5. Figure 10.6 is made easier to understand. Different simulation results may be obtained for execution of one thread and two threads, however, the same result is obtained even when it is repeatedly executed in each. On the other hand, in the case of execution with three or more threads, the behavior occurs in which simulation results are different each time it is executed.

$$r_0 = b - Ax_0; \quad \beta_0 = \frac{1}{(r_0, r_0)}; \quad p_0 = \beta_0 r_0$$

for $k = 0, 1, \cdots$ **until** $\|r_k\| \leq \varepsilon \|b\|$ **do**

$$a_k = \frac{1}{(p_k, Ap_k)}$$

$$x_{k+1} = x_k + \alpha_k p_k$$

$$r_{k+1} = r_k - \alpha_k Ap_k$$

$$\beta_{k+1} = \frac{1}{(r_{k+1}, r_{k+1})}$$

$$p_{k+1} = p_k + \beta_{k+1} r_{k+1}$$

end

Fig. 10.4 Conjugate gradient method [9]

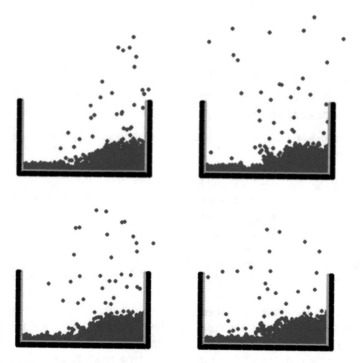

Fig. 10.5 Difference in simulation result (The four are at the same time and liquid may splash out of the container and sometimes it does not scatter)

10.2.2 Causes of Different Simulation Results

The difference in simulation results in parallel computing is caused by the usage of iterative solver first. In the case of the MPS method, it is a part of the conjugate gradient method. This is because subtle differences arise in the computation result when parallel computing is performed by the conjugate gradient method. In a time-

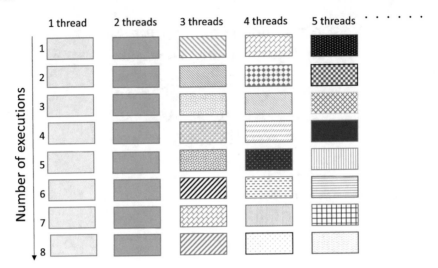

Differences in shade and pattern represent different simulation results.
When the number of threads becomes 3 or more, the result differs
with each simulation.

Fig. 10.6 A diagram that makes it easy to understand the difference in simulation results

lapse simulation, the small different behavior of some particles near the start may
have a big influence. The difference in execution results also affects the difference
in execution time, so it is a point to be aware of when tuning.

In what part of the conjugate gradient method the problem lurks, it is the result
part of the inner product of the two vectors across all threads. In the algorithm of Fig.
10.4, this is due to the fact that the result of the inner product operation $(\boldsymbol{p}_k, A\boldsymbol{p}_k)$
and $(\boldsymbol{r}_{k+1}, \boldsymbol{r}_{k+1})$ appearing in the formula for computing α_k and β_{k+1} is very slight
but different.

Assume that the number of threads is T, the thread number is i (range is $0 \sim$
$T - 1$), and the number of elements of the whole vector is n where n is divisible by
T. In addition, assume that each thread i is responsible for the elements from the
$i * n/T$ th to the $(i + 1) * n/T - 1$ th of the vector $\boldsymbol{a}$ and vector $\boldsymbol{b}$.

In the inner product computation, each thread computes the inner product of
partial vectors as follows:

$$s_i = \sum_{k=i*n/T}^{(i+1)*n/T-1} a_k \cdot b_k$$

After that, the sum of the inner product of the partial vectors of each thread is calculated.

$$s = \sum_{k=0}^{T-1} s_k$$

In the calculation of $s_0 + s_1$, the same result is obtained each time, but in the case of $s_0 + s_1 + s_2$, the calculation results may differ slightly from the calculation of $(s_0 + s_1) + s_2$ and $s_0 + (s_1 + s_2)$. In other words, in the addition of three or more numerical values, the calculation result varies depending on the order.

A calculation method that results in the same each time and has the smallest error is to rearrange in order of magnitude of absolute value and to add in decreasing order of absolute value. Although it is a method known as a countermeasure against a problem of remarkably poor convergence, sorting is required before addition, so overhead is big when it is done every time iteration.

10.2.3 Improvement Strategy

Not only is the simulation result different every time parallel execution, it becomes difficult to evaluate algorithms and implementation methods when the number of iterations until convergence changes at the iterative solving method for solving simultaneous linear equations and the execution time also changes accordingly. We will describe the countermeasures here.

10.2.3.1 When Always Executing with the Same Number of Threads

If the simulation is always performed fixed without changing the number of threads and the number of nodes, the countermeasure is relatively simple and the overhead is small. Normally, when calculating the sum between nodes or between threads, the calculation order is not defined. Therefore, when obtaining a sum, it is to obtain a sum in a predetermined order of the thread number and the node number. In the conjugate gradient method, we do that by inner product operation of the two vectors. For example, in the case of eight threads, we fix the order of operations that adds the values of individual threads as follows:

$$s = ((s_0 + s_1) + (s_2 + s_3)) + ((s_4 + s_5) + (s_6 + s_7))$$

If this is executed with at least the same number of threads, the results of the inner product operation of the vectors are exactly the same regardless of how many times they are executed, and the same simulation result can be obtained.

Some compilers and MPI libraries have extended functions that can fix the computation order and maintain repeatability when executed with the same number of threads or the same number of processes, so here is a brief introduction.

Intel Compiler

The Intel compiler's OpenMP runtime provides Intel extension environment variables that control runtime behavior. Specifically, by specifying 1 (default 0) for KMP_DETERMINISTIC_REDUCTION, the order of floating-point operations performed in the reduction directive clause of OpenMP is fixed and consistent floating results are obtained.

In the case of the Intel MPI library, by selecting the algorithms guaranteeing the operation order in the MPI collective operation control environment variables I_MPI_ADJUST_ALLREDUCE and I_MPI_ADJUST_REDUCE, the same computation result can be obtained.

Fujitsu Compiler

Fujitsu compiler can specify -Kordered_omp_reduction at compile time, this enables the order of operations executed by the reduction directive of OpenMP is fixed in order of thread number. When omitted, the -Knoordered_emp_reduction option is assumed to be specified and the operation order is not fixed.

In Fujitsu MPI library, by setting coll_base_reduce_commute_safe 1 which is one of the MCA parameters like mpiexec -mca coll_base_reduce_commute_safe 1, it is possible to fix the order of the reduction operation.

10.2.3.2 Countermeasures for Changing the Number of Threads

Basically, you need to write your own program in order to achieve the same result even if running with different number of threads. Specifically, the least common multiple of the number of threads considered to be executed is obtained, and the vector is divided by the least common multiple so as to carry out inner product calculation. For example, if the number of threads is scheduled to be 1, 2, 4, 8, 10, 16, 20, 40, divide the vector into 80 least.

In the case of 1, 2, 3, 4, 6, 12 threads (least common multiple is 12), the program code which obtains the same dot product value every time is shown in Fig. 10.7. It is impossible to deal with arbitrary number of threads in practice.

```
#define MIN(a, b)  (((a) < (b)) ? (a) : (b))

double ddot(double p[], double q[], int n)
{
  int i, j, n_12;
  double l_s, s[12];

  n_12 = (n - 1) / 12 + 1;

  #pragma omp parallel for private(j, l_s)
  for(i = 0; i < 12; i++){
    l_s = 0.0;
    for(j=i*n_12; j<MIN((i+1)*n_12,n); j++) l_s+=p[j]*q[j];
    s[i] = l_s;
  }
  return s[0] + s[1] + s[2] + s[3] + s[4] + s[5]+
         s[6] + s[7] + s[8] + s[9] + s[10] + s[11];
}
```

Fig. 10.7 An inner product operation routine that returns the same value each time (corresponding to the number of threads which is a divisor of 12)

10.2.4 Summary

Needless to say, it is very important to realize reproducible simulation also in parallel computation. In addition, it is not preferable to evaluate the performance that the number of times of convergence in the iterative solution part changes every time it is executed and the execution time of the whole simulation changes greatly. Therefore, it is important to determine in which part of the parallel computation the computation result will differ, and devises to minimize the influence on the execution speed while maintaining the reproducibility of the simulation.

10.3 High-Precision and High-Accuracy Calculation

Usually, floating-point operations on computers are done in binary64, i.e., double precision, and have approximately 16 significant decimal digits. However, there are unexpectedly many cases where even such high accuracy is insufficient. Here, the simplest solution would be to use higher precision, i.e., employ libraries for multiple-precision arithmetic beyond that of double. The idea itself seems obvious at first glance, but there is little support on software and hardware about it, and there is little consensus among venders. This chapter explains this situation and choices of libraries which are suitable for our purposes. We will assume that users have installed Ubuntu 16.04 on which to run examples.

10.3.1 Necessity for High-Precision and High-Accuracy Calculation

The floating-point formats were explained in Chap. 7. Whereas floating-point numbers are how computers handle real numbers, they do not satisfy distributive law and associative law. This is the origin of their numerical error. Most floating-point operations on computers today comply with the IEEE 754-2008 [10] standard. In particular, hardware usually implements binary32, i.e., single precision, and binary64, i.e., double-precision, floating-point numbers. Thus, arithmetic operations are very fast in these formats. It is possible to obtain about 7.5 Tflops in double-precision and 15 Tflops in single precision. Binary32 has 7 significant decimal digits, and binary64 has 16 significant decimal digits.

However, even these precisions have insufficient accuracy for some purposes. For them, IEEE 754-2008 defines binary128, i.e., quadruple precision, with 34 significant decimal digits. One might then wonder whether we need even more accuracy or is binary128 sufficient? The answer depends on the situation; in some cases, we do need more.

First, let us examine situations when lower precision calculations yield unexpected values, when the operations between floating-point numbers involve large errors, and when the effective digit is greatly reduced. If such problems cannot be avoided by some other means, then high precision calculations are needed. In following examples, we use eight-digit decimal numbers with an exponent range for -100 to 100 to clearly show the large errors of arithmetic operations.

- No matter how many times a small number, 1.0×10^{-9}, is added to 1.0;

$$1.0 + 1.0 \times 10^{-9} = 1.0$$

this sum becomes 1.0.
- Cancellation of significant digits: When subtracting numbers of approximately the same size

$$3.141592653 - 3.141592652 = 1.0 \times 10^{-9}$$

There are no errors in these operations, but the effective digits decrease to one digit! For example, when solving a quadratic equation for x,

$$ax^2 + bx + c = 0$$

where a, b, and c are real numbers and $b > 0$. The solution formula

$$x = \frac{-b \pm \sqrt{b^2 - 4ac}}{2a}$$

is not used as is. Instead, x_0, is first obtained as

$$x_0 = \frac{-b + \sqrt{b^2 - 4ac}}{2a}$$

However, x_1 is obtained as

$$x_1 = \frac{c}{ax_0}.$$

by using the relation of sum and product of roots (Vieta's formulas). When $|b| \sim \sqrt{b^2 - 4ac}$, the number of significant digits is reduced by cancellation.

- Overflow:

$$1.0^{100} \times 1.0^{100} = 1.0^{200}$$

This operation does not look problematic, but the result is not expressible in this format.

When an arithmetic operation exceeds the supported range, it may give an unexpected result. Normally, a library developed for numerical calculations will include algorithms that avoid operations with such problems. A more complicated case has already been shown in Chap. 7. That is, inversion of a matrix is not used when we solve a simultaneous linear equation on a computer. Instead, we perform LU factorization and then solve for the lower and upper triangle matrices. There are various situations in which it is a good idea to take the digit and exponent parts of large numbers [11, 12]. Examples are as follows:

- Long duration simulations, large-scale calculations in physics.
- Experimental mathematics, verification, and debugging. When a certain series should converge to zero, but a mathematical proof is not available, you can numerically observe whether convergence to zero progresses gradually with increasing accuracy. This then becomes evidence of convergence; e.g., such a procedure has been used when counting the zeroes of Riemann's ζ function [13].
- The Kahan summation algorithm, which sums sequences. It can be used to calculate sums to an accuracy that is twice the precision used [14].
- Subtracting a large number from another large number. When we evaluate Feynman integrals by using a Monte Carlo method, we end up having to subtract a large number from another large number. This operation induces cancellation of significant digits.
- In an iterative method such as BiCGStab, the number of iterations required for convergence decreases when a multiple-precision arithmetic operation is used [15].
- When solving a semidefinite programming problem with high-precision. When solving such a problem, Cholesky decomposition into symmetric matrices is performed many times. If Cholesky decomposition is performed in double-precision, only about eight significant decimal digits are correct in general. Numerical insta-

bility may occur when there are no inner points. In such a case, a high-precision arithmetic calculation is required [16, 17].

- Analyzing the asymmetric eigenvalue problem. This sort of problem is very difficult to solve in general, unlike the case of a symmetric matrix. High-accuracy calculations may be required.
- Verification of quantum electrodynamics (QED). QED is a very accurate physical model, but it does not take into account the effect of quantum chromodynamics (QCD). High-accuracy calculations are required when we want to determine the slight differences between QED and QCD predictions or between QED predictions and experimental results.
- Solving the ten high-precision calculation problems in contemporary mathematics first computed by Nick Trefethen in SIAM News 2002 [18].

There are many situations in which multiple-precision arithmetic becomes important. However, as we showed in Chap. 7, the distributive and associative laws are not strictly obeyed in floating-point arithmetic. A program written under the assumption that these rules strictly hold true for floating-point numbers may not work as expected (people who would want such a computing environment would not be interested in high-precision arithmetic in the first place).

10.3.2 On the Types, Accuracy, and Speed of Multiple Precision

"Multiple precision" is not a well-defined phrase. In fact, it is used to refer to (i) binary128 floating-point numbers, i.e., quadruple-precision numbers, which have approximately 32 decimal significant digits, (2) "double-double" numbers, which are concatenations of two double-precision numbers that have almost the same precision as binary128, or (3) multiple-precision numbers in which fractions can be extended arbitrarily. Binary128 is described in IEEE 754-2008. However, no commercially popular CPUs implement it in hardware; only software implementation is available. Thus, it is quite slow. "Double-double" numbers are realized in a tricky way, but their operations are the fastest among the three. Multiple precision yields most accuracy (but at the expense of speed) because it allows users to extend the length of fractions.

Multi-precision arithmetic is slow because the amount of computation it entails is drastically larger than in single precision. Moreover, most variations of it lack hardware support such as SSE or SIMD, and are made slower by the number of SIMD units in the CPU. For example, the software implementation of binary128 is about 100 times slower than the hardware implementation of double precision. Computations with very long numbers (e.g., ones with 1000 significant decimal digits) are even slower. As compensation, "double-double" precision has almost the same precision as binary128, and its operations can be performed as series of double-precision operations, meaning that it can benefit from hardware support such

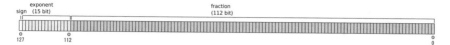

Fig. 10.8 See binary128 in the diagram. From Wikipedia

as SIMD. For that reason, double-double precision is *only* 10–20 times slower than double precision, and frequently used.

Nevertheless, multiple-precision arithmetic is very important; some numerically difficult problems can not be solved without it.

10.3.3 Binary128, or Quadruple Precision

Binary128 [10], or quadruple precision, is a floating-point number format with a 1-bit sign, 15-bit exponent width, and 112-bit fraction part, for 128 bits in total. The format is shown in Fig. 10.8. The first bit of the fraction part is 1. Thus, it can save 1 bit more and essentially uses 113 bits. It can express a decimal number up to $\log_{10}(2^{113}) = 34$ digits. Quadruple precision is supported by the GNU Compiler Collection (GCC) and Intel C/C++/Fortran compiler. Figure 10.9 shows a sample program in C that calculates π in double-precision and in quadruple precision and shows their difference.

Here, the double-precision result for π matches the actual value up to 15 digits. On the other hand, the quadruple precision calculation matches it up to the 34th digit; the difference is about 1.0×10^{-16}. The results are consistent with each other.

C or C++ has a big problem with regard to the language specification. C99 and C11 do not include quadruple precision; it is an extension specific to GCC. However, using __float128 is the right, even though it is not listed in the specifications.[4]

10.3.4 Double-Double Precision

Double-double precision, which has approximately 32 significant decimal digits and is a "cheaper" version of binary128 or quadruple precision, works by handling two

---

[4]When __float128 is defined in the standard C/C++, it is only necessary to switch __float128 with typedef. C/C++ are problematic in the size or interpretation of numbers; they may be different in different implementations or architectures, e.g., the "long double" can be either 80 bit (extended precision), IEEE 754 binary64 (double-precision), or IEEE 754 binary128 (quadruple precision). Moreover, "long double" means double-double by compilers on IBM Power processors. Even with the same 64-bit architecture, the data models such as LLP 64, LP 64, and ILP 64 are different; if two different binaries with the same program on the same machine and the same OS is compiled by two different compilers using different data models, they may give different results (segmentation fault usually occurs for unintended data model).

```
$ cat test.c
#include <quadmath.h>
#include <math.h>
#include <stdlib.h>
#include <stdio.h>
int main ()
{
    __float128 r, s, t;
    double _t;
    char buf[128];
    r = atanq(1.0q) * 4.0q;
    _t = atan(1.0d) * 4.0d;
    t = _t;
    quadmath_snprintf (buf, sizeof buf, "%+-#*.38Qe", r);
    printf("%s\n", buf);
    quadmath_snprintf (buf, sizeof buf, "%+-#*.38Qe", t);
    printf("%s\n", buf);
    s = r - t;
    quadmath_snprintf (buf, sizeof buf, "%+-#*.38Qe", s);
    printf("%s\n", buf);
}
$ gcc test.c -lquadmath ; ./a.out
+3.14159265358979323846264338327950279748e+00
+3.14159265358979311599796346854418516159e+00
+1.22464679914735317635888491926262295573e-16
```

Fig. 10.9 `__float128` sample program and execution in C

double-precision numbers. A double-double precision number a can be represented by using double-precision numbers a_{hi}, a_{lo}:

$$a = a_{hi} + a_{lo}$$

For this purpose, we make use of two algorithms called "quick-two-sum" and "two-prod". The former computes the exact sum of two double numbers [19]. The latter multiplies two double numbers exactly [20]. Using these algorithms as building blocks, it is possible to perform additions and multiplications of double-double numbers by using only double-precision operations. By using the C++ class created by Hida et al., double-double type can be used in programs in almost the same way as double-precision [21].

Recent CPUs implement the fused multiply add (FMA) instruction in their hardware.[5] When it is used, FMA reduces more than half of the double-precision operations involved in multiplication. Table 10.4 shows the number of double-precision

[5]FMA performs $a \times b + c$ in one clock, and it performs $a \times b + c$ exactly and rounds the result to double-precision. It is often used for inner product calculations and matrix–matrix multiplications. Why such hardware is implemented in recent CPUs is that since every instruction must be processed in one clock, both an adder and multiplier must exist in its arithmetic unit. The processor would stall if this were not the case. Implementing FMA on a CPU is a good way to these utilizing these two operators maximally, as it fills up the adder and multiplier in the arithmetic unit.

Table 10.4 Basic algorithm of double-precision arithmetic and number of double-precision arithmetic operations for addition and multiplication algorithms with double-precision using it

Algorithm	Op. count
Additions	20
Mulitiplications	24
Multiplications using FMA	10

operations for double-double arithmetic. According to the table, it is possible to perform double-double precision arithmetic using 10–24 double-precision operations.

Another advantage of this format is that we can make use of SIMD extensions such as SSE4 or AVX4.[6]

For example, on the implementation using NVIDIA C2050 [22], double-double matrix operations ran at about 16 Gflops, whereas genuine quadruple precision would have taken more than 100 times longer.

However, we cannot replace double-double precision instead of binary128 in all cases; it has some drawbacks as well. First, the precision of double-double arithmetic is lower by 3–4 decimal significant digits; the fraction part of double-double is 104 bits, while it is 113 bits in binary128. Second, the range of exponent part is smaller than in even double-precision.

Let us show another example of incompatibility of these two. The machine epsilon ε is the smallest value expressed as follows:

$$1 + \varepsilon \geq 1.$$

We can obtain approximation of a machine epsilon by following way; setting $\varepsilon = 1$. Then, calculate $\varepsilon + 1$. If this is larger than 1, divide ε by two. Repeat until $1 + \varepsilon$ becomes 1 in desired precision. Then, the previous value of ε is an approximation to the machine epsilon. In binary128 precision, ε is approximately 1×10^{-34}, while in double-double precision, surprisingly, it is 2.2×10^{-308}, which is the smallest number of the double precision but is not 1×10^{-32}.

10.3.5 GMP and MPFR Libraries that Can Deal with Fractions of Arbitrary Length

We will end this discussion by introducing the genuine multiple-precision libraries included in Ubuntu 16.04 packages: GMP [23], and MPFR [24].

One of the most important things in such as experimental mathematics is when the effective digit is quadruple precision at all or it can not be predicted. In this case, it is

---

[6]SSE4 and AVX4 stand for Streaming SIMD Extensions and Intel Advanced Vector Extensions, and they can perform operations such as double-precision numbers collectively with one instruction.

necessary to calculate an arbitrary length of the fraction part. GMP can be used for such purposes. It handles arithmetic operations of arbitrary length as well as rational numbers. GMP compares it with various high precision arithmetic libraries. It also belongs to the fastest category.

Unlike GMP, MPFR implements exponential functions, logarithmic functions, trigonometric functions as well as gamma and zeta functions. It has a rounding mode and can handle such as infinity, non a number (NaN), overflow as specified in IEEE 754. The complex computation library MPC [25] is based on MPFR and it gives meaning to such criteria and data.

Finally, there are the MPACK [26, 27] and BNCpack [28] libraries. These are linear algebra libraries such as BLAS and LAPACK. When we use arbitrary high-precision numbers in real applications, we rarely use FORTRAN77 or Fortran90 and we usually use C. However, still, it is very troublesome. If you use MPACK, you can use multiple-precision numbers in the same sense as "double" by C++. It is better to use a wrapper like this.

Exercises

1. Compile and run Alg1 and Alg2 in Sect. 10.1.5 in your own environment and take the same data as shown in Table 10.3. At that time, check the difference due to the difference in compiler options.
2. Compile the code in Fig. 10.6 and check that the numbers match for 1, 2, 3, 4, 6, and 12 threads. Also, make sure that there are errors in the number of threads, for other numbers.
3. By using a double-double precision library (apt-get install libqd-dev), compare the values of π calculated with double-double and double precision.
4. Obtain the inverse matrix (Rgetri) of the Hilbert matrix using MPACK [26, 27] with the MPFR library. Calculate the numerical error by: (1) taking the product of the Hilbert matrix and its inverse, (2) subtracting the unit matrix, and (3) calculating the 2-norm. Verify that the numerical error decreases as the precision is made larger.

References

1. IEEE Standard for Floating-Point Arithmetic, Std 754–2008 (2008)
2. S. Oishi, *Numerical Methods with Guaranteed Accuracy* (Corona-sya, 2000, Japanese)
3. E. Ramon, R. Moore, B. Kearfott, J. Michael, *Introduction to Interval Analysis* (Society for Industrial and Applied Mathematics, Cloud, 2009)
4. S. Oishi, S.M. Rump, Fast verification of solutions of matrix equations. Numer. Math. **90**(4), 755–773 (2002)
5. T. Ogita, S.M. Rump, S. Oishi, *Verified solution of linear systems without directed rounding, Technical Report 2005–04* (Waseda University, Tokyo, Japan, Advanced Research Institute for Science and Engineering, 2005)
6. N.J. Higham, *Accuracy and Stability of Numerical Algorithms*, 2nd edn. (SIAM Publications, Philadelphia, 2002)

7. T. Ogita, S.M. Rump, S. Oishi, Accurate sum and dot product. SIAM J. Sci. Comput. (SISC) **26**(6), 1955–1988 (2005)
8. S. Koshizuka, Y. Oka, Moving-particle semi-implicit method for fragmentation of incompressible fluid. Nuclear Sci. Eng. **123**, 421–434 (1996)
9. H. Togawa, *Conjugate Gradient Method* (Kyoiku Shuppan, 1977, in Japanese)
10. IEEE, IEEE standard for floating-point arithmetic, IEEE Std 754-2008, pp. 1–70 (2008)
11. D.H. Bailey, R. Barrio, J.M. Borwein, High precision computation: mathematical physics and dynamics. Appl. Math. Comput. **218**, 10106–10121 (2012)
12. D.H. Bailey, J.M. Borwein, High-precision arithmetic in mathematical physics. Mathematics **3**, 337–367 (2015)
13. G. Beliakov, Y. Matiyasevich, A parallel algorithm for calculation of large determinants with high accuracy for GPUs and MPI clusters. arXiv:1308.1536v2
14. N.J. Higham, *SIAM: Society for Industrial and Applied Mathematics*, 2nd edn. (2002)
15. H. Hasegawa, Utilizing the quadruple-precision floating-point arithmetic operation for the krylov subspace methods, in *Proceedings of the 8th SIAM Conference on Applied Linear Algebra*, vol. 25 (2012)
16. M. Nakata, B.J. Braams, K. Fujisawa, M. Fukuda, J.K. Percus, M. Yamashita, Z. Zhao, Variational calculation of second-order reduced density matrices by strong n-representability conditions and an accurate semidefinite programming solver. J. Chem. Phys. **128**, 164113 (2008)
17. H. Waki, M. Nakata, M. Muramatsu, Strange behaviors of interior-point methods for solving semidefinite programming problems in polynomial optimization. Comput. Opt. Appl. **53**, 823 (2012)
18. F. Bornemann, D. Laurie, S. Wagon, J. Waldvogel, *The SIAM 100-Digit Challenge: A Study in High-Accuracy Numerical Computing* (Society for Industrial and Applied Mathematics, SIAM, 2004)
19. D.E. Knuth, *Art of Computer Programming, Volume 2: Seminumerical Algorithms*, 3rd edn. (Addison-Wesley Professional, 1997)
20. T.J. Dekker, A floating-point technique for extending the available precision. Numerische Math. **18**, 224–242 (1971)
21. Y. Hida, X.S. Li, D.H. Bailey, *Library for double-double and quad-double arithmetic, Technical report* (Lawrence Berkeley National Laboratory, 2008)
22. M. Nakata, Y. Takao, S. Noda, R. Himeno, A fast implementation of matrix-matrix product in double-double precision on nvidia C2050 and application to semidefinite programming, in *Third International Conference on Networking and Computing (ICNC)* (2012)
23. T. Granlund, Gmp Development Team, *GNU MP 6.0 Multiple Precision Arithmetic Library* (Samurai Media Limited, United Kingdom, 2015)
24. L. Fousse, G. Hanrot, V. Lefevre, P. Pélissier, P. Zimmermann, MPFR: a multiple-precision binary floating-point library with correct rounding. ACM Trans. Math. Softw. **33**, 13 (2007)
25. A. Enge, M. Gastineau, P. Théveny, P. Zimmermann, mpc—a library for multiprecision complex arithmetic with exact rounding, INRIA, 1.0.3 edn., Feb 2015
26. M. Nakata, MPACK, RIKEN, 0.8.0 edn. (2012)
27. M. Nakata, Mpack0.6.7: a high precision linear algebra library. Appl. Math. **2110** (2011, In Japanese)
28. T. Koya, BNCpack, 0.7 edn. (Shizuoka Institute of Science and Technology, 2011)
29. B.N. Parlett, *The Symmetric Eigenvalue Problem (Classics in Applied Mathematics)* (Society for Industrial Mathematics, 1987)

Index

Printed in the United States
By Bookmasters